COMPUTERIZED MULTIPLE INPUT CHROMATOGRAPHY

ELLIS HORWOOD SERIES IN ANALYTICAL CHEMISTRY

Series Editors: Dr R. A. CHALMERS and Dr MARY MASSON, University of Aberdeen
Consultant Editor: Prof. J. N. MILLER, Loughborough University of Technology

S. Alegret	Developments in Solvent Extraction
S. Allenmark	Chromatographic Enantioseparation—Methods and Applications
G.E. Baiulescu, P. Dumitrescu & P.Gh. Zugravescu	Sampling
H. Barańska, A. Łabudzińska & J. Terpiński	Laser Raman Spectrometry
G.I. Bekov & V.S. Letokhov	Laser Resonant Photoionization Spectroscopy for Trace Analysis
K. Beyermann	Organic Trace Analysis
O. Budevsky	Foundations of Chemical Analysis
J. Buffle	Complexation Reactions in Aquatic Systems: An Analytical Approach
D.T. Burns, A. Townshend & A.G. Catchpole	Inorganic Reaction Chemistry Vol. 1: Systematic Chemical Separation
D.T. Burns, A. Townshend & A.H. Carter	Inorganic Reaction Chemistry: Vol. 2: Reactions of the Elements and their Compounds: Part A: Alkali Metals to Nitrogen, Part B: Osmium to Zirconium
J. Churáček	New Trends in the Theory & Instrumentation of Selected Analytical Methods
E. Constantin, A. Schnell & A. Pepe	Mass Spectrometry
R. Czoch & A. Francik	Instrumental Effects in Homodyne Electron Paramagnetic Resonance Spectrometers
T.E. Edmonds	Interfacing Analytical Instrumentation with Microcomputers
Z. Galus	Fundamentals of Electrochemical Analysis, Second Edition
S. Görög	Steroid Analysis in the Pharmaceutical Industry
T. S. Harrison	Handbook of Analytical Control of Iron and Steel Production
J.P. Hart	Electroanalysis of Biologically Important Compounds
T.F. Hartley	Computerized Quality Control: Programs for the Analytical Laboratory
Saad S.M. Hassan	Organic Analysis using Atomic Absorption Spectrometry
M.H. Ho	Analytical Methods in Forensic Chemistry
Z. Holzbecher, L. Diviš, M. Král, L. Šůcha & F. Vláčil	Handbook of Organic Reagents in Inorganic Chemistry
A. Hulanicki	Reactions of Acids and Bases in Analytical Chemistry
David Huskins	Electrical and Magnetic Methods in On-line Process Analysis
David Huskins	Optical Methods in On-line Process Analysis
David Huskins	Quality Measuring Instruments in On-line Process Analysis
J. Inczédy	Analytical Applications of Complex Equilibria
M. Kaljurand & E. Küllik	Computerized Multiple Input Chromatography
S. Kotrlý & L. Šůcha	Handbook of Chemical Equilibria in Analytical Chemistry
J. Kragten	Atlas of Metal-ligand Equilibria in Aqueous Solution
A.M. Krstulović	Quantitative Analysis of Catecholamines and Related Compounds
F.J. Krug & E.A.G. Zagotto	Flow Injection Analysis in Agriculture and Environmental Science
V. Linek, V. Vacek, J. Sinkule & P. Beneš	Measurement of Oxygen by Membrane-covered Probes
C. Liteanu, E. Hopîrtean & R. A. Chalmers	Titrimetric Analytical Chemistry
C. Liteanu & I. Rîcǎ	Statistical Theory and Methodology of Trace Analysis
Z. Marczenko	Separation and Spectrophotometric Determination of Elements
M. Meloun, J. Havel & E. Högfeldt	Computation of Solution Equilibria
M. Meloun, J. Militky & M. Forina	Chemometrics in Instrumental Analysis: Solved Problems for IBM PC
O. Mikeš	Laboratory Handbook of Chromatographic and Allied Methods
J.C. Miller & J.N. Miller	Statistics for Analytical Chemistry, Second Edition
J.N. Miller	Fluorescence Spectroscopy
J.N. Miller	Modern Analytical Chemistry
J. Minczewski, J. Chwastowska & R. Dybczyński	Separation and Preconcentration Methods in Inorganic Trace Analysis
T.T. Orlovsky	Chromatographic Adsorption Analysis
D. Pérez-Bendito & M. Silva	Kinetic Methods in Analytical Chemistry
B. Ravindranath	Principles and Practice of Chromatography
V. Sedivec & J. Flek	Handbook of Analysis of Organic Solvents
O. Shpigun & Yu. A. Zolotov	Ion Chromatography in Water Analysis
R.M. Smith	Derivatization for High Pressure Liquid Chromatography
R.V. Smith	Handbook of Biopharmaceutic Analysis
K.R. Spurny	Physical and Chemical Characterization of Individual Airborne Particles
K. Štulík & V. Pacáková	Electroanalytical Measurements in Flowing Liquids
J. Tölgyessy & E.H. Klehr	Nuclear Environmental Chemical Analysis
J. Tölgyessy & M. Kyrš	Radioanalytical Chemistry, Volumes I & II
J. Urbanski, *et al.*	Handbook of Analysis of Synthetic Polymers and Plastics
M. Valcárcel & M.D. Luque de Castro	Flow-Injection Analysis: Principles and Applications
C. Vandecasteele	Activation Analysis with Charged Particles
F. Vydra, K. Štulík & E. Juláková	Electrochemical Stripping Analysis
N. G. West	Practical Environmental Analysis using X-ray Fluorescence Spectrometry
J. Zupan	Computer-supported Spectroscopic Databases
J. Zýka	Instrumentation in Analytical Chemistry

COMPUTERIZED MULTIPLE INPUT CHROMATOGRAPHY

M. KALJURAND, Ph.D.
Senior Researcher

and

E. KULLIK, Ph.D.
Head, Department of Instrumental Analysis
both of Institute of Chemistry
Estonia SSR Academy of Sciences, Tallin, USSR

Edited by
W. A. J. BRYCE and R. A. CHALMERS
University of Aberdeen

ELLIS HORWOOD LIMITED
Publishers · Chichester

Halsted Press: a division of
JOHN WILEY & SONS
New York · Chichester · Brisbane · Toronto

First published in 1989 by
ELLIS HORWOOD LIMITED
Market Cross House, Cooper Street,
Chichester, West Sussex, PO19 1EB, England
The publisher's colophon is reproduced from James Gillison's drawing of the ancient Market Cross, Chichester.

Distributors:

Australia and New Zealand:
JACARANDA WILEY LIMITED
GPO Box 859, Brisbane, Queensland 4001, Australia

Canada:
JOHN WILEY & SONS CANADA LIMITED
22 Worcester Road, Rexdale, Ontario, Canada

Europe and Africa:
JOHN WILEY & SONS LIMITED
Baffins Lane, Chichester, West Sussex, England

North and South America and the rest of the world:
Halsted Press: a division of
JOHN WILEY & SONS
605 Third Avenue, New York, NY 10158, USA

South-East Asia
JOHN WILEY & SONS (SEA) PTE LIMITED
37 Jalan Pemimpin # 05–04
Block B, Union Industrial Building, Singapore 2057

Indian Subcontinent
WILEY EASTERN LIMITED
4835/24 Ansari Road
Daryaganj, New Delhi 110002, India

© 1989 M. Kaljurand and E. Kullik/Ellis Horwood Limited

British Library Cataloguing in Publication Data
Kaljurand, M. (Mitikel) *1945–*
Computerised multiple input chromatography.
1. Chemical analysis. Applications of computer systems
I. Title II. Kullik, E. (Endel)), *1929–*
543.′0028′5

Library of Congress Card No. 88-29300

ISBN 0-7458-0120-X (Ellis Horwood Limited)
ISBN 0-470-21228-4 (Halsted Press)

Printed in Great Britain by Hartnolls, Bodmin

COPYRIGHT NOTICE
All Rights Reserved. No part of this publication may be reproduced, stored in a retrieval system, or transmitted, in any form or by any means, electronic, mechanical, photocopying, recording or otherwise, without the permission of Ellis Horwood Limited, Market Cross House, Cooper Street, Chichester, West Sussex, England.

Contents

Foreword 9

Preface 13

1 INTRODUCTION 15
   1.1 Input–output relationship of a chromatograph 15
   1.2 Exploration of some input functions 17
       1.2.1 Step and impulse techniques 18
       1.2.2 Multiple-input techniques 19
   1.3 Random input and correlation function 22
   1.4 Convolution and cross-correlation computation 26
   1.5 Summary of input functions 26
   References 27

2 CORRELATION CHROMATOGRAPHY 28
   2.1 Multiplex measurements 28
       2.1.1 Principles of multiplex measurement 28
       2.1.2 Examples of multiplex measurements 32
   2.2 Theory of correlation chromatography (CC) 37
       2.2.1 Input–output relationship and a circulant matrix . 37
       2.2.2 Decorrelation of the output sequence 40
       2.2.3 Signal-to-noise ratio in correlation chromatography, and the multiplex advantage 42
       2.2.4 Non-stationary input 48
           2.2.4.1 Random variations in the sampled quantities of the flow 49
           2.2.4.2 Systematic variations in the sampled quantities of the flow 51
       2.2.5 Validation of the formulae for correlation noise . 53
   2.3 Comparison of CC with other methods for the measurement of low-concentration samples 54

      2.3.1 Injecting large samples 54
      2.3.2 Sample concentration 55
  References 57

3 TWO-DIMENSIONAL MEASUREMENTS INVOLVING CHROMATOGRAPHY 59
  3.1 Introduction 59
  3.2 Time-resolved chromatography of non-stationary concentration flows 60
      3.2.1 The sampling theorem 60
      3.2.2 Equi-interval sampling and high-speed separations. 63
      3.2.3 Analysis of time-varying gas and liquid flows by correlation chromatography 68
      3.2.4 Ensemble correlation and stroboscopic sampling 70
  3.3 Two-dimensional separation by column-switching 73
      3.3.1 Column-switching in liquid chromatography 76
      3.3.2 Column-switching in GC 79
  3.4 Amount of information obtained with two-dimensional separation systems 84
      3.4.1 Uncertainty and information 84
      3.4.2 Chromatographic peak capacity and the amount of information obtained by column chromatography 87
      3.4.3 Amount of information obtained with a two-dimensional system 90
  3.5 Selectivity tuning 92
  3.6 Presentation of two-dimensional chromatograms 94
      3.6.1 Isometric projection 94
      3.6.2 Slices of two-dimensional chromatograms 95
      3.6.3 Contour plots 96
  References 97

4 INSTRUMENTATION 100
  4.1 Basic set-up of multiple-input computerized chromatography 100
  4.2 Sample source 101
      4.2.1 Individual sample sources. 101
      4.2.2 Pyrolysis reactors 105
          4.2.2.1 Sample preparation for pyrolysis reactors 106
          4.2.2.2 Temperature control in pyrolysis reactors 107
  4.3 Input systems. 118
      4.3.1 The effect of physical and chemical factors on the flow 119
      4.3.2 Pneumatic flow switches. 122
      4.3.3 Flow switches based on fluidic elements. 129
      4.3.4 Mechanical and solenoid valves 131
      4.3.5 Autosamplers and robot sampling 133
      4.3.6 Summary of input systems 134
  4.4 Chromatographs: demands of multiple input 134

4.5 Computers: demands of multiple-injection chromatography 136
References 138

5 **APPLICATION OF MULTIPLE-INPUT COMPUTERIZED CHROMATOGRAPHY** 139
   5.1 Characterization of gases evolved from polymers. 139
      5.1.1 Various approaches to polymer characterization 142
         5.1.1.1 Pyrolysis mass-spectrometry 142
         5.1.1.2 Pyrolysis infrared spectrometry 147
         5.1.1.3 Methods of thermal analysis 150
         5.1.1.4 Historical remarks: development of the pyrolysis method 151
      5.1.2 Analysis and applied pyrolysis gas chromatography 153
         5.1.2.1 Single-input pyrolysis gas chromatography; conventional method 156
         5.1.2.2 Multiple pyrolysis gas chromatography 157
      5.1.3 Thermochromatography 162
         5.1.3.1 Scope of thermochromatography. 163
         5.1.3.2 Experimental conditions in thermochromatography. . . . 165
         5.1.3.3 Earlier applications of thermochromatography 171
         5.1.3.4 Rubbers. 173
         5.1.3.5 High-temperature resistant polymers: the role of flame retardants 178
         5.1.3.6 Desorption of gases from solids. 184
         5.1.3.7 Study of the influence of catalysts on polymer thermal stability 185
         5.1.3.8 Correlation gas chromatography in characterization of high-temperature resistant polymers. 185
   5.2 Applications of correlation chromatography 186
      5.2.1 Historical remarks 186
      5.2.2 Application of CC in gas analysis 189
      5.2.3 Application of CC in water analysis 192
   References 195

Appendix 1: Some matrix algebra 198

Appendix 2: A feedback shift register and the fast Hadamard transform 202

Appendix 3: Software for the correlation chromatography 209
      Appendix 3.1: Generation of permutations 209
      Appendix 3.2: Experiment control program. 212
      Appendix 3.3: Decorrelation program 213

Appendix 4: Contour plot BASIC subroutine 215
      Appendix references 216

Notation 217

Abbreviations 219

Index 220

Foreword

Today computers have penetrated everywhere into our life and challenge humans in a wide area of intellectual endeavours from playing chess to the checking of the English spelling in this manuscript with the "Wordstar" and "Spellstar" text editors. Chromatography is no exception and was one of the first branches of analytical chemistry where computers were extensively used [1].

Now computers are playing an important role in chromatographic instrumentation.

The application of computers to chromatography (as in the whole of analytical chemistry) can be divided into three branches: (1) instrument control, (2) data processing and (3) chemometrics. The first two branches were extensively developed in the beginning of the seventies and approximately four hundred papers on controlling chromatographs by microcomputers, data processing and interfacing chromatographs to computers were published. These early results have been documented in several books [2–6]. Later the number of papers in these areas decreased significantly. One possible reason for this is that the manufacturers of chromatographic instruments made every effort to construct new microprocessor-controlled chromatographs with built-in calculation of peak areas and retention times. These instruments are good enough but sometimes disappoint old-fashioned chromatographers: nothing can be changed or reconstructed on the chromatograph without the modification of the control program. These instruments are valuable tools for routine application, but for chromatographic research (other than on application of computers) they are of less interest.

Chemometrics is the "application of mathematical and statistical methods to chemical measurements" [7]. However, this definition is too wide and lacks precise boundaries and content. A distinctive feature of chemometrics is that it deals with problems in analytical chemistry that cannot be solved without a computer. It is the manipulation of a large amount of data by computers to reveal information hidden behind the numbers, or the control of experiments by a computer in such a way that the problem-solving and the decision-making are based on the results of processing large data sets. Chemometrics uses methods of information theory, automatic control, multivariate statistics, factor analysis, optimization, pattern recognition and experimental design.

As it is understood in contemporary analytical chemistry, chemometrics is the processing of the data obtained experimentally (e.g. pattern recognition or plotting of multidimensional data). This book describes how a chromatographer can control his instrument by computer, i.e. the accent is put on instrument control and active participation in the experiment. It also describes new types of experiments in chromatography, where computers are closely linked to the equipment, i.e. these experiments cannot be performed without a computer. A well-known example outside chromatography is computer tomography and Fourier spectroscopy. Equivalent experiments in chromatography are possible. This is a very exciting area of investigation, with many unresolved problems. Thus, this book is mainly directed towards the graduate student or young researcher searching for his own project.

The aim of a measurement is to obtain new knowledge. This means that computer-based chromatography should feature increased selectivity and sensitivity, the first being associated with the separation process and the second with the detection process. These goals can be achieved by replacing the traditional microsyringe injection of a sample into the chromatograph. One reason for introducing a sample by microsyringe is the need to obtain an understandable chromatogram with an interpretable location of peaks, and a baseline. This restriction is removed in computerized chromatography. The computer can introduce a sample in a very complicated way by following predetermined time sequences or a mathematical function. A detector signal that is not fully interpretable by the human can be converted into intelligible form by extensive computation. The amount of new knowledge that computerized chromatography is able to give can be expressed in terms of information theory and it follows that the *amount* of information is larger than that acquired in traditional chromatography.

This book describes improvements in the separation process by 'multidimensional' chromatography as well as in the detection process by 'multiplex'' chromatography. These objectives are realized by use of equipment capable of multiple injections of a sample under computer control. It justifies dealing with both topics within one book. The term 'multi' is a keyword of this book.

The diversity of problems dealt with in this book may lead to confusion but it reflects only the diversity of ideas that are in use in modern computer-aided analytical chemistry. In fact the text should be useful for a specialist working in several different fields: spectroscopists, chemometricians, computer software developers and especially chemists who study polymer stability. It contains many examples of the application of computerized multiple input chromatography in polymer degradation studies, which was the speciality of the book's authors for many years.

Moscow   KARL I. SAKODYNSKII

REFERENCES

[1] *J. Chromatog. Sci.* 1969, **7**, 12.
[2] A. L. Gurevich, L. A. Kolomytsev and L. A. Rusinov, *Automatization of Chemographic Information Processing*, (in Russian), Energiya, Moscow, 1973.
[3] J. K. Foreman and P. B. Stockwell, *Automatic Chemical Analysis*, Ellis Horwood, Chichester, 1975.

[4] E. Küllik, M. Kaljurand and M. Koel, *Application of Computers to Gas Chromatography*, (in Russian), Nauka, Moscow, 1978.
[5] A. L. Gurevich, L. A. Rusinov and N. A. Sygaev, *Automatic Chromatographic Analysis*, (in Russian), Khimiya, Leningrad, 1980.
[6] R. E. Kaiser and A. J. Rackstraw, *Computer Chromatography*, Vol. 1., Hüthig Verlag, Heidelberg, 1983.
[7] B. Kowalski, *Anal. Chem.*, 1980, **52**, 112R.

Preface

Our aim was not to cover all chemometric applications of chromatography but only those linked to multiple injections and problems associated with them. We have attempted to write this book so that no prior knowledge of the terms and concepts used in chemometrics is necessary. Where unavoidable, basic terms are introduced. However, matrix notation is extensively used because of its convenience. Only a knowledge of elementary matrix algebra is necessary to understand our conclusions. We do not go further than matrix multiplication. This, however, does not mean that the book is intended to be easy reading. Some of the derivations in multiplex chromatography are quite lengthy. Also we do not assume that readers have any knowledge of computer architecture but a knowledge of programming will be useful.

As multiplex and multidimensional chromatography are quite different methods the corresponding chapters can be read separately and it is not necessary to read the whole book to understand the respective parts.

We especially thank our colleagues from the Laboratory of Instrumental Analysis of the Institute of Chemistry, Tallinn, Drs. M. Koel and T. Sõmer and Prof. C. Cramers from Eindhoven Technical University, The Netherlands, for reading and commenting on various parts of the manuscript.

We thank sincerely Prof. J. Lindberg and Dr. M. Elomaa of Helsinki University and Dr. M. Romanov of the Institute of Plastic Materials, Moscow, for valuable suggestions about the possible applications of multiple input chromatography to polymer degradation research and for supplying samples for study. Some of the results of these studies are used as examples in this book.

In addition we thank Riina Süld for reading and correcting the manuscript.

Finally we are grateful to the host of authors from whose works we have learned. They are too many to name, but they are acknowledged in the references at the end of the chapters.

Tallinn,
January 1987

MIHKEL KALJURAND
ENDEL KÜLLIK

1
Introduction

1.1 INPUT–OUTPUT RELATIONSHIP OF A CHROMATOGRAPH

A chromatograph may be considered to consist of three different parts: an injector, a column, and a detector. The input signal to a chromatograph is the sample mixture of substances. The components of the mixture pass through the column at different speeds and cause the detector to respond to their appearance at the end of the column, by giving an electrical output signal. As the aim of chromatography is the physical separation of a mixture the input to the chromatograph should be a concentration impulse of short duration (theoretically infinitely narrow). Owing to diffusion and mass transfer processes in the column, an initially infinitely narrow band spreads in the column and the detector registers not an infinitely narrow band but a band with a width that is a function of time, $h(t)$. This function has a finite duration and in the case of a mixture $h(t)$ is the sum of all the responses to the mixture components. In automatic control theory this function is called a system impulse response function.

What will the detector response be if the input is not an infinitely narrow impulse? Let us first assume that n concentration impulses of short duration, Δt, are injected into a column and that the amount of sample injected at the moment $i\Delta t$ is $x(i\Delta t)\Delta t$. Here $x(i\Delta t)$ is the rate at which the substance is supplied to the chromatograph inlet, and can be considered constant during a short period of time Δt. The overall detector response at the time t is the sum of individual responses:

$$y(t) = \sum_{i=0}^{n} x(i\Delta t) h(t - i\Delta t) \Delta t \tag{1.1}$$

For $n = 2$, $y(t)$ is presented in Fig. 1.1. If $\Delta t \to 0$ and $n \to \infty$, then the sum in Eq. (1.1) approaches the convolution integral well-known in several fields of applied sciences [1]:

$$y(t) = \int_{-\infty}^{\infty} x(\tau) h(t - \tau) d\tau = \int_{-\infty}^{\infty} x(t - \tau) h(\tau) d\tau \tag{1.2}$$

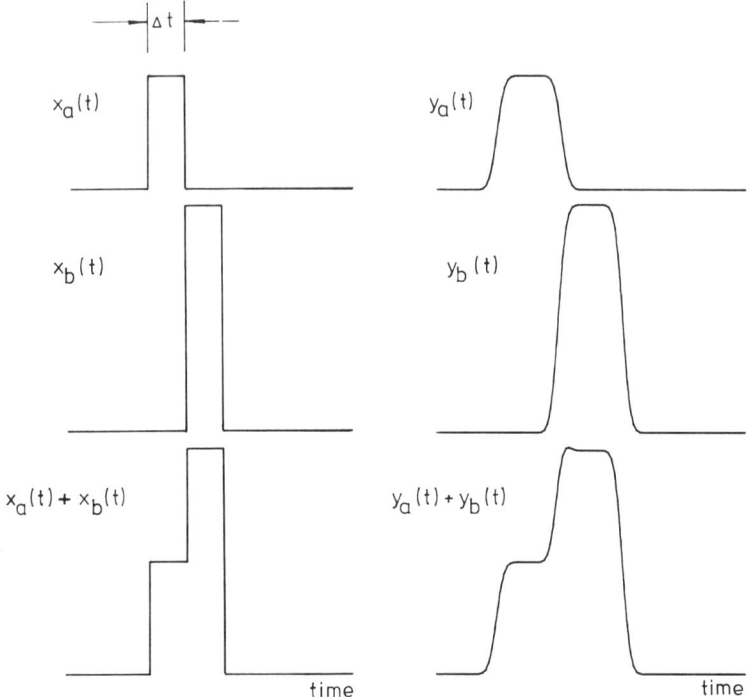

Fig. 1.1 — Linearity of the impulse response, $x_a(t)$, $x_b(t)$ for different input functions, $y_a(t)$, $y_b(t)$, and corresponding detector responses.

The latter equation has been obtained by changing the variables. Equation (1.2) expresses the most general relationship between input and output. It enables us to calculate the detector response to any input function $x(t)$. In the case of a narrow impulse integration is performed over a very short period of time, say Δt, and $x(t)$ is approximately a constant, x_0, so integration leads to $y(t) = x_0 h(t)$, i.e. the expected result of an impulse response. A narrow single injection is the predominating input in chromatography because it leads (not always, of course) to separate component bands at the end of the column and to easily interpretable chromatograms in the form of a sequence of peaks. This form of injection is called a single injection. All other injection forms are called multiple or complex injections or input functions.

The description of the chromatographic process by means of the convolution integral regards a chromatograph as a linear system [1]. This means that in calculating a detector response individual impulses influence the chromatograph independently and an overall response is the sum of all individual responses as was already assumed in deriving Eq. (1.1). This equation is a statement of the superposition principle that linear systems must satisfy. It is generally valid in chromatography, or at least the chromatographer tries to perform measurements in such a way that all non-linearities can be neglected.

If the assumption about the system linearity is not valid then the description of the system becomes complicated. The superposition principle does not hold any more and the relationship between system input, $x(t)$, and output, $y(t)$, will be described by the Volterra series [1]. The series is rather complicated and we do not give it here. The Volterra series has been successfully used in the non-linear analysis of noise in nuclear magnetic spectroscopy [2]. Non-linearity is not at all rare in chromatography. Column overloading by the sample, interaction between sample components in the chromatographic column, or between sample components and the liquid phase are well-known sources of non-linearities in chromatography. The flame photometric detector has a non-linear response. Using the Volterra series approach for analysis of non-linearities in chromatography may give some interesting results. The analysis involves a large number of computations and they are not well suited for the microcomputers that have so far been successfully used in chromatography. A new generation of microcomputers, the so-called personal computers, have a megabyte of memory and some are equipped with mathematical processors performing very fast floating point computations. These computers may be well suited for the analysis of chromatographic non-linearities.

1.2 EXPLORATION OF SOME INPUT FUNCTIONS

So far, the input–output relationship in chromatography has been discussed in terms of the automatic control theory that regards a system as a 'black box' with an unknown response function, $h(t)$. By exciting the 'black box' with the input function, $x(t)$, and recording the output signal, $y(t)$, it is possible to get some information about $h(t)$. In chromatography, the situation is a little bit different. The theory of chromatography gives some hints about the impulse response function $h(t)$. Although a number of quite complicated expressions for $h(t)$ are available in the chromatographic literature, the Gaussian curve is the most popular. Several chromatographic band-spreading theories lead to the Gaussian curve when the separation time approaches infinity. For l sample components the detector response function, the chromatogram, is as follows:

$$h(t) = \sum_{i=1}^{l} \frac{f_i A_i}{\sqrt{2\pi}\,\sigma_i} \exp\left[-\frac{(t - t_{R_i})^2}{2\sigma_i^2}\right] \tag{1.3}$$

For the ith peak, A_i is the ith sample component amount, t_{R_i} is the location of the peak maximum, f_i is the detector response factor and σ_i is the peak standard deviation related to the peak half-width $w_{1/2}$ by $w_{1/2} = 2\sigma\sqrt{2\ln 2} = 2.3548\sigma$.

Generally, the aim of a chromatographic measurement is the determination of A_i and t_{R_i} and sometimes also σ_i. As mentioned above, a single narrow input is the best for this purpose because then A_i, t_{R_i} and σ_i can be directly measured from the chromatogram. However, very early on Reilley et al. [3] pointed out that various complex input functions can also be utilized in chromatography. Let us consider some input functions and their corresponding output signals. Various input functions are presented in Fig. 1.2.

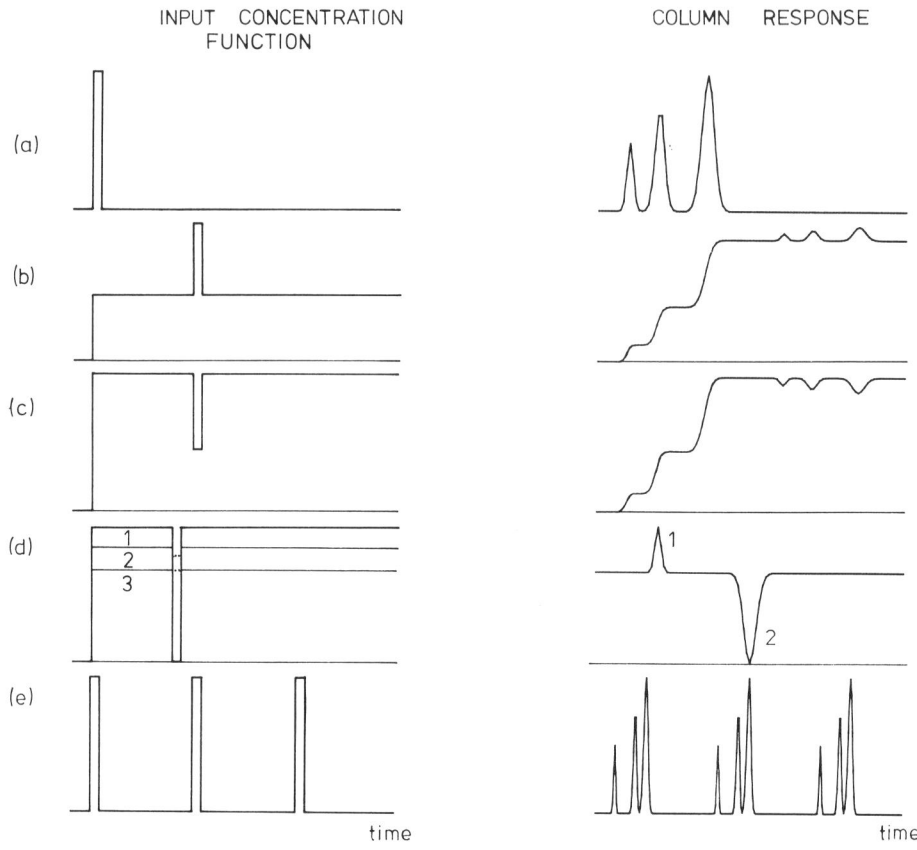

Fig. 1.2 — Responses of the chromatographic column to various input functions.

1.2.1 Step and impulse techniques

A single impulse input is presented in Fig. 1.2a and, as follows from Eq. (1.2), it results in a common chromatogram. If the single impulse is superimposed by a step it leads to the function in Fig. 1.2b. The output is a constant baseline with a common chromatogram superimposed on it. This technique enables us to eliminate adsorption effects in the column. Irreversible adsorption on the interfacial gas–liquid and liquid–solid surfaces is a common phenomenon in gas chromatography and contributes to the retention time [4]. By use of this type of input function, the column is theoretically pre-equilibrated and adsorption sites are saturated, allowing partition effects to be observed without interference from adsorption.

A fast stepwise rise of concentration from zero to some finite value in the column inlet is smoothed out in the detector response, by which we can follow a characteristic stepwise incease in the response function. The number of steps is equal to that of components and the step height is proportional to the component amount. It follows from Eq. (1.2) that the response function for a component in the case of a step input is equal to $\frac{1}{2}A\{1 + \text{erf}[(t - t_R)/\sqrt{2}\sigma]\}$ where erf(t) is the error integral [5].

In 'vacancy' chromatography [6], a single pulse of the pure carrier gas is made into the sample stream. The corresponding input function is shown in Fig. 1.2c. The response function represents the reversed chromatogram with negative peaks. The peak areas are proportional to the concentrations of the corresponding flow components.

A further development of vacancy chromatography is differential chromatography [7] where instead of pure carrier gas, samples with a similar composition are injected. The method enables us to measure small differences in sample composition in the presence of a large amount of the matrix compound (e.g. trace species in air or water). Let us consider the following example. Let us assume that we have the simplest gas chromatograph with a thermal conductivity detector. First, the reference stream flows through the first path of the detector, an injector, a column, and the second path of the detector. What does the detector output look like if an analytical sample is taken from a flow of interest having a composition similar to that of the reference flow and injected into the chromatograph? It is clear that the flow components for which the concentrations are the same in the analytical and reference flow give no signal in the detector at all. If the component concentration in the analytical flow is higher than that in the reference flow, then the output response will be a peak in a positive direction and in the opposite case, the peak is reversed. So, this technique should be useful in process monitoring. The idea of differential chromatography is shown in Fig. 1.2d. In this figure, numbers denote components and the corresponding band geometrical height is proportional to concentration. In the impulse input to the three-component mixture the concentration of the third component is unchanged, the concentration of the first component is increased and that of the second component is decreased compared with the original flow.

Vacancy chromatography has some advantages over common chromatography. Injection of pure carrier gas is sometimes simpler than that of the analytical sample. It is possible to work in the non-linear region of the sorption isotherm, the active adsorption centres are saturated, and thus the peaks are more symmetrical. This technique has been applied to the separation of nucleic acid bases on an exchange column, with a liquid mobile phase [8]. The technique was used to mask unwanted components. However, the UV detector performance was affected by a high background signal. This may be true for other detectors as well and is a disadvantage of vacancy chromatography.

Iteration chromatography [7] is another variant of differential chromatography where component concentrations in the analytical flow are changed continuously during measurement so that the detector signal disappears. Then the analytical and reference flows should be identical and the concentration of components in the analytical flow is known from the reference flow. The advantage of iteration chromatography is that variations in performance with time (carrier gas flow-rate, solvent bleeding, temperature and electronics drift) do not affect the results because the detector performs as a null instrument.

1.2.2 Multiple-input techniques

Let us consider the repetitive impulse input such as that illustrated in Fig. 1.2e. We call this input equi-interval sampling because the time interval between single impulses is constant. This sequence is useful for studying flows which are continuously changing in composition (e.g. monitoring the gases evolved from a chemical reaction). In Section 3.2.1 we will show that the rate of variation in the flow must be related to the time interval between impulses for an adequate description of the changes in flow.

Equi-interval sampling can be most conveniently achieved under computer control. However, some pioneering work on performing this sampling manually was done two decades ago [9,10]. These authors attempted to overcome the limitation of batch separation of continuous flow components in common chromatography, by taking advantage of the structure of a particular chromatogram. Knowing in advance the gas hold-up time and the distances between peaks it was not necessary to wait for complete elution of a particular chromatogram from a particular injection before injecting the next sample, and this allowed them to shorten the intervals between samplings.

The simplest way to realize equi-interval sampling in reaction gas chromatography is to introduce a reagent at the top of the column, filled with catalyst, and periodically stop the carrier gas flow. Products formed when the flow is stopped are swept onto the column and separated there when the flow is switched on again. A typical stopped-flow chromatogram consists of sharp product peaks superimposed on the reaction chromatogram that would be obtained under continuous flow conditions (Fig. 1.3). The stopped-flow technique was proposed by Phillips et al. [11] and developed by Katsanos and Tsiatsios [12].

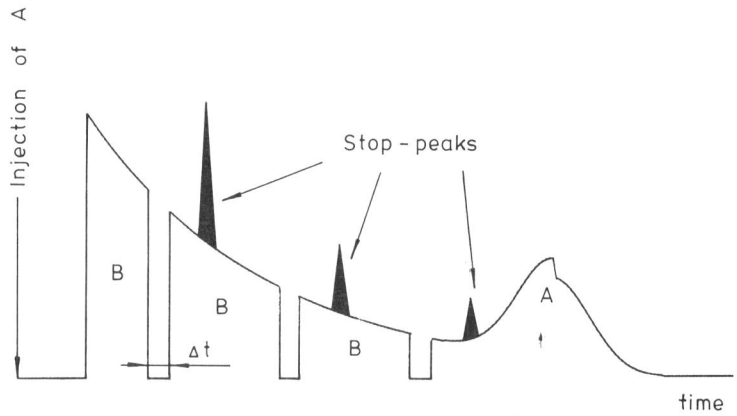

Fig. 1.3 — Stopped flow chromatogram of an A→B type reaction; Δt, stop time.

Frequency modulation chromatography is a modification of equi-interval sampling. It was introduced by Villalanti et al. [13]. The corresponding input function resembles the equi-interval pulse sequence (which may also be called a constant frequency input). Frequency modulation is obtained by slightly shifting each input impulse from the position dictated by constant frequency sampling. Before each pulse, a random number is generated to determine the magnitude and sign of deviation from the carrier frequency for the next injection time. The output signal, $y(t)$, is demodulated by subtracting it from its value after a delay of one cycle of the carrier signal: $y'(t) = y(t+\Delta t/2) - y(t-\Delta t/2)$ where $y'(t)$ is the demodulated signal and Δt is the duration of a cycle. This output signal is later decorrelated from the input signal to get a chromatogram. The decorrelation operation will be considered in Section 1.3. Frequency modulated chromatography enables us to maintain the solute concentration almost constant during an experiment. This means that the partition coefficients are constant even in working in the non-linear region of an isotherm. This property, in turn, may be advantageously used when working with large amounts of sample.

Possibilities of the computerized input systems are not restricted to equi-interval sampling. A computer can generate whatever sequence of time intervals is necessary to realize some particular input form. For example, in two-dimensional chromatography a computer can analyse the first detector signal and control the sampling to the second column. This technique of multiple sampling will be discussed more thoroughly in Section 3.3.1.

A sinusoidal input offers the possibility of continuous analysis of flow components. The steady-state response to this input is

$$y(t) = A_0 + \sum_i A_i \exp[-2(\pi f \sigma_i)^2] \sin[2\pi f(t - t_{R_i})]$$

where f is the input wave frequency, and A_0 is the steady-state response [3]. It follows from this formula that $y(t)$ is also a wave with the same frequency as the input wave but the phase is shifted by $2\pi f t_{R_i}$ and the response wave amplitude is attenuated by the exponential factor, the value of which depends on the frequency f and the component-band standard deviation σ_i. This permits determination of one component in the presence of another without interference by the latter. Let us assume that the standard deviation of the component of interest is σ_1 and the second component has a standard deviation σ_2. The second component sine wave amplitude is attenuated relative to the first by a factor of $\exp(z)$ if the input wave frequency is

$$f = \frac{1}{2\pi\sigma_2} \sqrt{\frac{2[z - \ln(A_1/A_2)]}{1 - (\sigma_1/\sigma_2)^2}}$$

The reader can derive this easily himself. If, for example, $\sigma_1 = 1$ sec, $\sigma_2 = 3$ sec, $A_1 = A_2$ and $z = 4.6$, ($e^z = 100$), then $f = 0.17$ Hz. On the other hand, if $\sigma_1 = 1$ sec, $\sigma_2 = 1.1$ sec, $A_2 = 10A_1$ and $z = 4.6$ then $f = 1.29$ Hz. All the components eluting later than the second one are suppressed more because the σ_i values are larger than those of the second component. This method has, however, a limited use because only the first evolving component can be determined in such a way.

By using more than one detector it is possible to detect the other components as well if the input function is a sine wave [14–16]. The method enables a truly continuous monitoring of flow component changes. The continuous chromatograph consists of several short chromatographic columns connected in series with non-destructive detectors. The flow of the carrier and sample is continuously fed to the chromatograph inlet at constant rate and modulated according to the sine wave input function. At the detectors, sinusoidal outputs are also observed. However, the phases of the waves of each detector output are shifted relative to the input function phase and these phase shifts are functions of the sample concentration and the retention time. Thus, in order to determine the sample composition the flow phase (or amplitude) is measured at the detectors. A functional relationship between phase shift and amount of component can easily be derived by using simple trigonometric identities. The problem reduces to the solving of a set of linear equations. Component retention times must be known beforehand.

The number of detectors must be one less than the number of components. The latter condition determines the possible area of application of this technique; the monitoring of gas flows with simple components. The detector is the most expensive and complicated part of the chromatograph and the need for a large number of detectors severely limits the use of the sinusoidal input. It has not been developed since the initial work [14–16].

The contents of this section may seem to be lumbered with descriptions of methods that have remained outside the main stream of chromatography despite their several valuable advantages. The application of computers may change the situation crucially and the value of the complex input techniques described may become fully utilized. Differential chromatography and frequency modulated chromatography, in particular, have good perspectives in trace analysis when they are coupled with modern computer-based noise-suppression techniques.

In the last section of this chapter we consider the solving of the convolution integral for $h(t)$ and comment on a very special input function: a random impulse sequence. This input function is the most exciting amongst the family of complex input functions. It has a definite advantage over the single input function, viz. the multiplex advantage. The random input function enables us to build up a nice theory and use effective computation algorithms for evaluation of chromatograms; several interesting new phenomena are associated with this input function in chromatography.

1.3 RANDOM INPUT AND CORRELATION FUNCTION

It is possible to evaluate the convolution integral in Eq. (1.2) in a closed form only for some simple input functions. Solving Eq. (1.2) numerically requires dealing with a set of linear equations of high order. Generally, this is not a simple task because of computer memory and speed limitations. Moreover, the system tends to be ill-conditioned, i.e. small errors in the matrix elements cause considerable variations in the solution. Random input functions considerably simplify the finding of a solution and permit us to achieve another important goal: suppression of the detector noise. Noise suppression cannot be obtained with a single injection to the chromatograph and this is the reason why interest in this unusual input form has considerably increased in the last few years.

Let us consider the convolution integral but now assuming that some noise, $r(t)$,

Random input and correlation function

is superimposed by the detector signal. In the case of additive noise the input–output relationship is

$$y(t) = \int_{-\infty}^{\infty} x(t-\tau)h(\tau)d\tau + r(t) \tag{1.4}$$

The cross correlation of two real functions, $x(t)$ and $y(t)$, is defined as [17]:

$$C_{x,y}(u) = \lim_{t \to \infty} \frac{1}{2t}\int_{-t}^{t} x(\tau-u)y(\tau)d\tau \tag{1.5}$$

It is similar to the convolution integral except that the component $x(t)$ is simply displaced to $y(t)$ without reversal. Cross correlation is easier to imagine than convolution but its properties are less simple. For example, a change of variables does not lead to the same function, i.e. the correlation operation is not a commutative operation as is that of convolution. The correlation function $C_{x,y}(u)$, measures similarity between two functions, $x(t)$ and $y(t)$. If $x(t) = y(t)$, then Eq. (1.5) determines an autocorrelation function. Substituting $y(t)$ from Eq. (1.4) into Eq. (1.5) we get:

$$C_{x,y}(u) = \lim_{t \to \infty} \frac{1}{2t}\int_{-t}^{t} x(\tau-u)d\tau \int_{-\infty}^{\infty} x(\tau-\lambda)h(\lambda)d\lambda$$

$$+ \lim_{t \to \infty} \frac{1}{2t}\int_{-t}^{t} x(\tau-u)r(\tau)d\tau \tag{1.6}$$

The second term in Eq. (1.6) is zero if the experiment time approaches infinity (i.e. $t \to \infty$) and functions $x(t)$ and $y(t)$ are independent. The function $r(t)$ is the experimental noise and frequently has a normal distribution. If the input is, for example, also a random function with a normal distribution ('white noise'), then the cross-correlation function of $x(t)$ and $r(t)$ is indeed zero for all displacement values, which means that the detector noise is suppressed. So, by changing the concentration at the chromatograph inlet in such a way that changes do not correlate with the detector noise it is possible to measure input–output cross correlation without noise. In practice, of course, experiment time is not infinite and some noise remains. In Section 2.2.3 we give an exact formula for the attenuation of the detector noise.

The random input has another advantage in that in this case the cross-correlation function coincides with the chromatogram. This can easily be proved by changing the order of integration in Eq. (1.6). Then we get the following result (known as the Wiener–Lee relation [18]):

$$C_{x,y}(u) = \int_{-\infty}^{\infty} h(\tau) C_{x,x}(u-\tau) d\tau \tag{1.7}$$

where $C_{x,x}(t)$ is the input autocorrelation function. It is known from statistics (see [18]) that for white noise $C_{x,x}(t) = C_0 \delta(t)$ where C_0 is a constant and $\delta(t)$ is the delta function. The delta function is the unit impulse with the following properties:

$$\delta(t) = \begin{cases} \infty\,; t = 0 \\ 0; t \neq 0 \end{cases} \int_{-\infty}^{\infty} \delta(t-u) h(t) dt = h(u) \tag{1.8}$$

It follows from these properties that integration of a function multiplied by the delta function results in the function value at the point zero and from this property and Eq. (1.7) it immediately follows that $C_{x,y}(u) = h(u) C_0$ in the case of the white noise input. A sample recording of white noise is shown in Fig. 1.4.

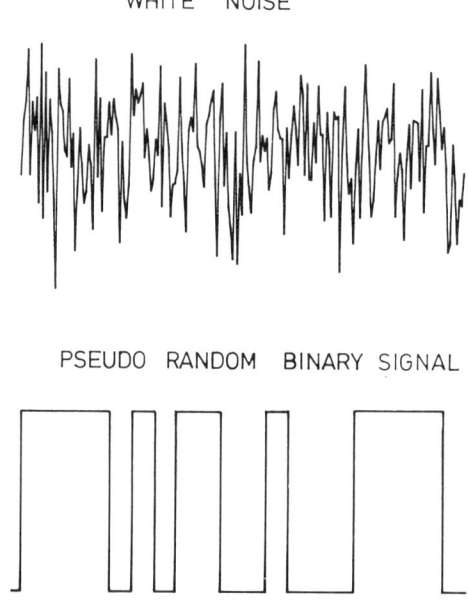

Fig. 1.4 — Pseudo random binary noise and white noise.

The white noise input is an idealized function of infinite length. Moreover, white noise contains all frequencies, i.e. its spectrum is flat (this justifies its name, white noise). Real random input functions have a finite length in time and are an approximation to white noise. Approximations of white noise may be generated by computer [18]. For chromatography, a pseudo-random binary noise input seems to be the best. The pseudo-random binary noise is generated according to the pseudo-random binary sequence (PRBS) with elements equal to $+1$ and -1. This noise type has only two levels (say, C and 0) and in our case, C means concentration level. The pseudo-random binary noise input to the chromatograph means that the computer generates a PRBS element and injects a sample into the chromatograph if the PRBS element is positive, otherwise it injects pure carrier gas. A typical example of PRBS noise is shown in Fig. 1.4. Experimentally it is very easy to obtain this input by computer-controlled valves. At the same time it is difficult to imagine how to realize experimentally the input function of the white noise type shown in Fig. 1.4.

We consider the generation of PRBS in Appendix 2. Here it is necessary to point out that PRBS can be generated only by $2^q - 1$ elements, where q is an integer. So, PRB sequences with 3, 7, 15, 31, 63, 127 ... elements are possible and that is sometimes considered a disadvantage of this noise type, but in fact it does not create large problems. If the sequence length is big enough, the autocorrelation function of the PRBS noise is a good approximation to the $\delta(t)$ function and cross correlation of input and output leads to the chromatogram according to Eq. (1.7). The width of the decorrelated chromatogram peak is, however, somewhat wider than in the case of a single infinitely narrow input impulse because of the finite width of the PRBS noise autocorrelation function peak. This dispersion (the chromatogram peak second moment), σ_T^2, was computed by Smit et al. [19] and is given by

$$\sigma_T^2 = \sigma^2 + (\Delta t)^2/6 \qquad (1.9)$$

where σ^2 has its common meaning of a Gaussian peak variance. If the cross-correlation function in Eq. (1.5) is computed by a digital computer and Δt is taken equal to the digitization interval, then according to the sampling theorem (see Section 3.2.1) $\sigma \geq 2\Delta t$ and it follows from Eq. (1.9) that the increase in the peak width is about 2%. Hence, the PRBS noise is a really good approximation to white noise.

A correlation experiment in chromatography involves several steps.

1. Computer generates a PRBS and controls the flow switching valves according to that sequence, thus producing a random input into the chromatograph.
2. Computer records the detector signal in its memory.
3. Computer computes the cross-correlation function between input and output [according to Eq. (1.5), or in some other way]. This function is equivalent to a single-input chromatogram except that the second moment has been increased by a small amount according to Eq. (1.9) and the high-frequency noise level has been decreased.

A purely random sequence of pulses is also in use as an input function in chromatography [20]. In this input form, the time interval between pulses is the random variable. The autocorrelation function, $C_{x,x}(t)$, approximates the $\delta(t)$ function, as do all the random input functions. This input function gives more freedom in choosing the length of the experiment. As already mentioned, a PRBS can have only a fixed number of elements, equal to $2^q - 1$. The purely random sequence has no such restriction. The algorithms for generating the purely random sequence and the PRBS are of the same level of sophistication. However, using the PRBS enables us to calculate a chromatogram by a more effective algorithm than is possible by using a purely random signal. We will discuss this in the next section.

1.4 CONVOLUTION AND CROSS-CORRELATION COMPUTATION

The computations involved when a complex input is used are worthy of mention. It is possible to evaluate the convolution integral in a closed form only for some simple input functions. If it is not possible to compute a chromatogram analytically, then it should be done numerically by a digital computer. This leads to the problem of inverting a matrix of order n^2 where n is the number of measurements. Usually, $n >$ 100–1000, which is a large number even for a microcomputer.

Taking advantage of the modern algorithms developed in numerical analysis it is possible to evaluate the convolution integral by the Fourier transform. The Fourier transform is a well-known mathematical operation which has a useful property: the Fourier transform of the convolution of two functions is equal to the product of multiplication of their Fourier transforms [5]. Thus, denoting the Fourier transform operation by $FT\{\}$, according to this property we get from Eq. (1.2) that $FT\{y(t)\} = FT\{x(t)\}FT\{h(t)\}$ and the chromatogram can be found from $h(t) = FT^{-1}\{FT\{y(t)\}/FT\{x(t)\}\}$ where $FT^{-1}\{\}$ denotes the inverse transform. The Fourier transform can be performed by the fast algorithm developed by Cooley and Tukey [21]. A computer program (with detailed description) is available [22]. The cross correlation function may be computed by use of this algorithm.

When the input function is PRBS noise, a chromatogram can be obtained by a direct cross correlation of the input with the detector output. In numerical computations, the input function reduces to a sequence of ones and zeros and this simplifies the computation by Eq. (1.5). Computation reduces to multiplication of the detector output digitized signal by an input matrix with a special structure, having elements equal to only -1 and $+1$. The algorithms for this computation are very easy to program even in machine code because no multiplications or divisions are involved (see Section 2.2.2).

The properties of the PRBS allow us to speed up computation of the chromatogram even more. Using a special rearrangement of the digitized output signal elements it is possible to compute the chromatogram by performing the fast Hadamard transform of the detector output signal. Programming the algorithm is a bit complicated and requires some knowledge but the transform gives an almost immediate result, especially when programmed in machine code. We describe this algorithm in more detail in Appendix 2.

1.5 SUMMARY OF INPUT FUNCTIONS

In everyday practice, a single narrow concentration pulse input is probably the best to use. However, for specific problems one of the special methods mentioned may have an advantage.

The content of this chapter can be reviewed as follows. The input function to the chromatograph, usually a narrow pulse, can take various forms. The detector output can be calculated by the convolution integral, Eq. (1.2). Two complex input function forms are of great interest: equi-interval sampling (Fig. 1.2e) for studying time-varying processes by chromatography and a random input for work with noisy detector signals. Neither case can be handled well by conventional chromatography.

The appearance of several complex input techniques in chromatography is due to the variety of the problems solved by chromatography. All complex and multiple input methods described in this chapter are summarized in Table 1.1.

Table 1.1 — Summary of input techniques in chromatography

No.	Input technique	Purpose
1	Step	reducing irreversible adsorption
2	Vacancy	simplification of sampling
3	Differential	measurement of deviations from the present flow
4	Iteration	elimination of the variations in system performance
5	Equi-interval	reaction kinetics studies and two-dimensional separations
6	Stopped-flow	reaction studies
7	Frequency modulated	working in the non-linear region of the solute isotherm
8	Sinusoidal	monitoring of the time-varying flow composition
9	Random	decreasing the detection limit

The list is not complete and the appearance of new, perhaps computerized, input techniques can confidently be expected.

REFERENCES

[1] P. Z. Marmarelis and V. Z. Marmarelis, *Analysis of Physiological Systems: The White-Noise Approach*, Plenum Press, New York, 1978.
[2] B. Blümich and D. Ziessow, *J. Chem. Phys.*, 1983, **78**, 1059.
[3] C. N. Reilley, G. P. Hildebrand and J. W. Ashley, *Anal. Chem.*, 1962, **34**, 1198.
[4] V. G. Berezkin, *Gas–Liquid–Solid Chromatography*, p. 112. Khimiya, Moscow, 1986, (in Russian).
[5] G. Korn and M. Korn, *Mathematical Handbook for Scientists and Engineers*, p. 720. McGraw-Hill, New York, 1961.
[6] A. A. Zhukhovitskii and N. M. Turkeltaub, *Dokl. Akad. Nauk USSR*, 1961, **143**, 646.
[7] A. A. Zhukhovitskii, *Zh. Analit. Khim*, 1972, **27**, 971.
[8] R. P. W. Scott, C. G. Scott and P. Kucera, in *Advances in Chromatography 1971*, A. Zlatkis (ed.), p. 219, Dept. Chemistry, Houston, 1971.
[9] D. Macnaughtan, Jr. and L. B. Rogers, *Anal. Chem.*, 1971, **43**, 822.
[10] H. R. Murdock, Jr., *Anal. Chem.*, 1970, **42**, 687.
[11] C. G. S. Phillips, A. J. Hart-Davis, R. G. L. Saul and J. Wormald, *J. Gas Chromatog.*, 1967, **5**, 424.
[12] N. A. Katsanos and A. Tsiatsios, *J. Chromatog.*, 1981, **213**, 15.
[13] D. C. Villalanti, M. F. Burke and J. B. Phillips, *Anal. Chem.*, 1979, **51**, 2222.
[14] S. Hiratsuka and A. Ichikawa, *Bull. Chem. Soc. Japan*, 1967, **40**, 2303.
[15] D. E. Carter and G. L. Esterson, *Ind. Eng. Chem. Fundam.*, 1970, **9**, 661.
[16] D. Obst, *J. Chromatog.*, 1968, **32**, 8.
[17] G. Horlick and G. M. Hieftje, in *Contemporary Topics in Analytical and Clinical Chemistry*, Vol. 3, D. Hercules, G. M. Hieftje, L. R. Snyder and M. Evenson (eds.), p. 154. Plenum Press, New York, 1978.
[18] G. A. Korn, *Random-Process Simulation and Measurements*, McGraw-Hill, New York, 1966.
[19] H. C. Smit, R. P. J. Duursma and H. Steigstra, *Anal. Chim. Acta*, 1981, **133**, 283.
[20] J. B. Phillips, D. Luu, J. Pawliszyn and G. C. Carle, *Anal. Chem.*, 1985, **57**, 2779.
[21] J. W. Cooley and J. W. Tukey, *Math. Comput.*, 1965, **19**, 297.
[22] E. O. Brigham, *The Fast Fourier Transform*, Prentice-Hall, Englewood Cliffs, NJ, 1978.

2
Correlation chromatography

Correlation chromatography (CC) utilizes the most sophisticated of all the inputs: the random input. As shown in the previous chapter, the random input is, in a sense, equivalent to a single input but has the advantage of producing a lower noise level in the chromatogram. This phenomenon is known as a multiplex advantage. In this chapter, we will consider the random input from a theoretical and practical point of view. CC utilizes a very general measurement principle applicable to every case where the output signal is a linear combination of input signals, i.e. the input–output relationship is given by the convolution integral, Eq. (1.2). We will consider several examples of applying the multiplex measurement principle in various branches of science. This should clarify the idea and ideology of CC and show its place among multiplex measurement systems. We will also show that although attractive from a theoretical point of view, practical application of CC encounters some difficulties and requires the solution of several problems that are still under investigation, restricting the wider application of this method.

2.1 MULTIPLEX MEASUREMENTS

2.1.1 Principles of multiplex measurement

Measurement methods (or experimental arrangements) are called multiplex if the output from the method (or system) is registered as a linear combination of the required results [1]. To get a 'traditional' output, the set of measurements from the multiplex instrument is transformed by a mathematical operation into a common form. Thus, performing a multiplex experiment requires rearrangement (or at least reorganization) of a conventional experiment and the performing of computations. Combination of sophisticated equipment and computational skills leads to various improvements in scope and performance, called the multiplex advantage.

Let us consider a simple example of weighing four unknown masses, m_1, m_2, m_3 and m_4, by a beam balance, with an error e_i [2]. Let us assume that e_i is a random variable with a zero mean and variance equal to σ^2. If all the objects are weighed separately, the ith balance reading is $y_i = m_i + e_i$. The estimate of m_i is y_i, with standard deviation equal to σ. On the other hand, let us suppose that the objects are weighed simultaneously by loading them in a certain order on both the pans of the balance. We would get the following set of balance readings:

$$y_1 = m_1 + m_2 + m_3 + m_4 + e_1$$
$$y_2 = m_1 - m_2 + m_3 - m_4 + e_2$$
$$y_3 = m_1 + m_2 - m_3 - m_4 + e_3 \quad (2.1)$$
$$y_4 = m_1 - m_2 - m_3 + m_4 + e_4$$

where all the objects are loaded on, say, the right-hand pan for the first weighing, and in the second weighing two objects are placed on the left-hand pan and two on the right, and so on. This set of equations can easily be solved for each m_i, e.g. for estimation of m_1' we get

$$m_1' = \tfrac{1}{4}(y_1 + y_2 + y_3 + y_4) = m_1 + \tfrac{1}{4}(e_1 + e_2 + e_3 + e_4)$$

It follows from elementary statistics that the variance of ce, where c is a constant, is $c^2\sigma^2$ and the variance of the sum of independent random variables is the sum of individual variances. Therefore, the variance of m_1' (and the other masses) is $4\sigma^2/16 = \sigma^2/4$. Weighing the objects simultaneously has reduced the variance of an unknown-weight estimation by a factor of 4.

This example has all the features of a multiplex measurement. A conventional weighing experiment was rearranged, the result was obtained by computation, and the precision was improved over that for separate weighing of the objects. Though trivial in the case of a small number of measurements, the computations became complicated in the case of a larger number of measurements and before the 'computer age' multiplex measurement methods were only of theoretical interest. The weighing design described above was proposed by Yates as early as 1935 [3]. Multiplex optical spectroscopy was discovered by Fellgett in the early 1950s [4]. The multiplex measurement principle, however, had been used far earlier by Fizeau [5] and Michelson [6]. The inventions of these great physicists formed the basis for modern Fourier-transform spectroscopy, the most widely used multiplex method.

How to design a multiplex experiment in chromatography? Some hints were given already in Section 1.3 while describing the application of the random input. Here we give a simplified physical model to aid in the understanding of the elements of correlation chromatography.

Suppose we have two flows: a sample flow and a carrier flow. A valve directs either the sample flow or the carrier flow to the chromatographic column. Let the valve position be determined by a pseudo-random binary sequence (PRBS). An example of the input sequence is $X = \{1,1,1,0,1,0,0\}$. Thus, during one PRBS cycle the valve is in the injecting position four times (denoted in the input sequence by ones) and in the non-injecting position three times (denoted by zeros). Let us also assume that the injection time, Δt, is relatively short, i.e. not long enough to cause extra-column band broadening. The detector output signal, Y, is digitized, after the interval Δt and the chromatogram we obtain is also in digitized form, say, $\boldsymbol{H} = \{0,0,1,0,0,2,0\}$. The number of chromatogram elements is chosen to be equal to that of PRBS elements. This is the requirement of CC and can easily be fulfilled by adding zeros to the chromatogram elements or by choosing a PRBS of proper length. If the PRBS and the chromatogram have n elements, the detector output has $2n-1$ elements. This is evident because all single-injection chromatograms are shifted relative to the first one in time and the contribution of the last injection begins from the nth element of the detector output and results in a chromatogram with n elements. Thus, the optimized CC experiment (the number of PRBS and chromatogram elements is equal) lasts twice as long as the elution of a single-injection chromatogram.

Let us assume, however, that two PRBS cycles are repeated during one CC experiment and that from the beginning of the nth element of the detector signal, all single-injection chromatograms give a contribution to each detector output element. The detector output constructed from single-injection chromatograms is presented in Fig. 2.1. Ignoring the first seven elements of the detector output, we can write the next seven as follows:

$$\begin{aligned} Y_8 &= H_1 + H_7 + H_6 + H_4 + e_1 \\ Y_9 &= H_2 + H_1 + H_7 + H_5 + e_2 \\ Y_{10} &= H_3 + H_2 + H_1 + H_6 + e_3 \\ Y_{11} &= H_4 + H_3 + H_2 + H_7 + e_4 \\ Y_{12} &= H_5 + H_4 + H_3 + H_1 + e_5 \\ Y_{13} &= H_6 + H_5 + H_4 + H_2 + e_6 \\ Y_{14} &= H_7 + H_6 + H_5 + H_3 + e_7 \end{aligned} \qquad (2.2)$$

where e_i is the detector noise element. Estimation of the chromatogram element H_i' can easily be obtained by solving the system of linear equations (2.2). For H_1',

$$H_1' = \tfrac{1}{4}(Y_8 + Y_9 + Y_{10} - Y_{11} + Y_{12} - Y_{13} - Y_{14}) = H_1 + (\Sigma_i e_i)$$

Therefore the variance of H_1' (and also of $H_2'-H_7'$) is $\sigma^2/16$ where σ is the detector-noise standard deviation. In the case of a single injection, $H_i' = Y_i$ and the variance of H_i' is σ^2. Again, as in the case of weighing the four masses, the noise level on the chromatogram decreases as compared to the single injection of the sample when using CC. Thus, CC is a multiplex experiment. The decrease in noise level is proportional to the square of the length of the input sequence.

From a linear-algebra point of view we can say that the relationship between the digitized chromatogram (or masses) and the detector output (or balance readings) is in the form of a system of linear equations that can be conveniently written as a matrix equation, $Y = XH$, where X is an $n \times n$ matrix. This equation has a simple solution: $H = X^{-1}Y$, where X^{-1} is the inverse of X. Finding X^{-1} and thus finding H is not an easy task if X has no special structured features. In our examples of weighing masses and chromatograms the elements of the matrix are 1 and 0. In Fourier-transform (FT)–infrared (FT–IR), FT–nuclear magnetic resonance (FT–NMR), FT–ion cyclotron resonance, and FT–ion mobility spectroscopy, the matrix elements are trigonometric functions of the type $\exp(2\pi i k l/n)$, where $i = \sqrt{-1}$ and k, l and n are integers. Inversion of this matrix took a long time before Cooley and Tukey discovered their famous fast Fourier-transform (FFT) algorithm [7] in 1965. The

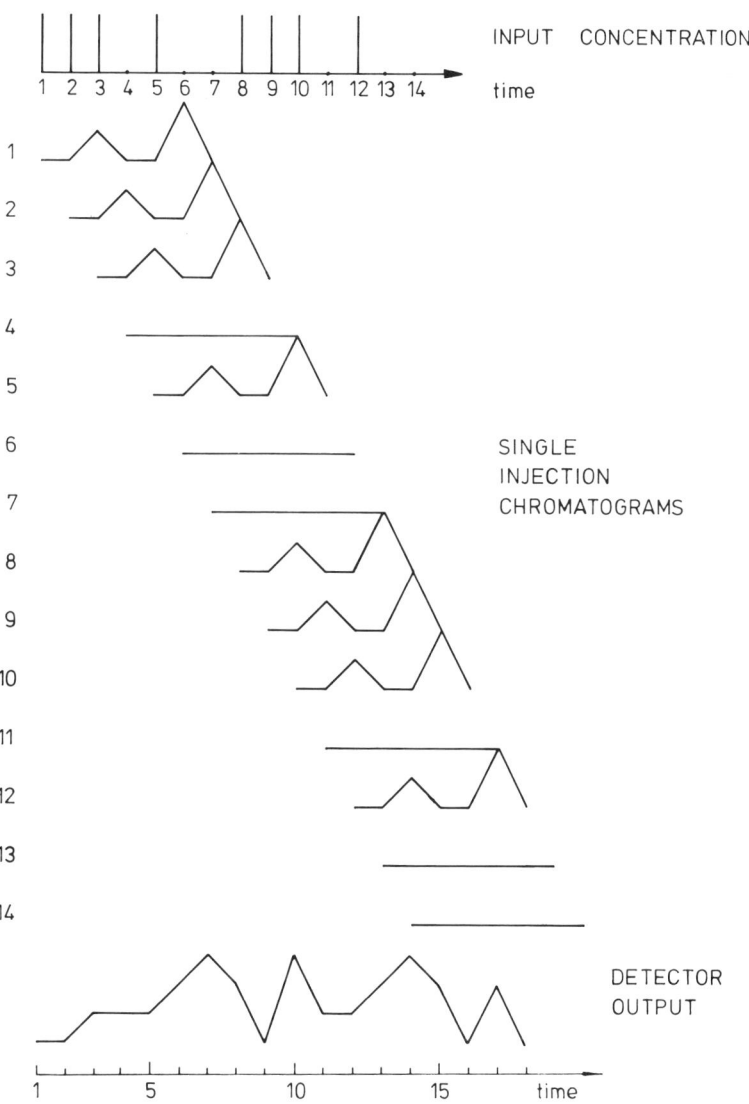

Fig. 2.1 — Construction of the output in correlation chromatography.

utilization of this algorithm in microprocessors, invented at the same time, ensured the break-through of Fourier spectroscopy.

Fourier-transform spectroscopy is the most widely used multiplex measurement system. Several instrument companies produce equipment for both FT–NMR and FT–IR. The method is in routine use and well documented in the literature [8,9]. The other forms of FT spectroscopy are less well known and their applications are still being investigated.

This chapter is concerned with matrices having elements of 0 and 1. The reason for this is the nature of the experiments under consideration. It is difficult to imagine a weighing experiment that results in a matrix with elements that are trigonometric functions. Multiplex methods that have matrices with elements of 1 and 0 are all described in the same mathematical terms despite the different physical nature of the measurements. Moreover, owing to the imperfections in equipment, several interesting phenomena (e.g. 'ghost' peaks) are common in these methods. In Section 2.1.2 we describe a few multiplex methods. Some of them have been referred to as Hadamard-transform (HT) spectroscopic methods. This is due to the fact that a matrix can be frequently converted into a special form first considered by the French mathematician Jacques Hadamard. Hadamard matrices have analytical expressions for the inverse matrix. The multiplication of a vector by the Hadamard matrix can be done by a fast algorithm similar to the fast Fourier-transform. We describe the Hadamard matrices in Appendix 2.

2.1.2 Examples of multiplex measurements

Examples and an exhaustive description of multiplex methods used in physics are given in [1] and those used in chemistry in [10,11]. The latter sources use the term 'transform techniques' instead of 'multiplex methods'. This is justified in the sense that multiplex methods lead to some kind of matrix multiplication or transformation of output data. Most of the multiple measurements can be described in terms of automatic control theory, regarding the object to be studied as a 'black box', and the aim is to find the impulse response reaction of the 'black box' (Fig. 2.2a). Theoretically the input–output relationship is given by the convolution integral, Eq. (1.2). Instead of a single-impulse input excitation of the system, a multiple (mainly PRBS, stochastic, or sinusoidal) excitation of the object is done in transform techniques. An analogy with chromatography is straightforward. In chromatography, the impulse response of the system is a chromatogram, in other fields, the meaning of the system input response follows from the nature of the system. In the following we shall describe some of the multiplex methods to demonstrate the generality of the multiplex principle.

Time-of-flight spectra are useful for characterization of particle speed distribution in molecular beams. The equipment involves a beam chopper and a counter, in a high vaccum. The chopper is a rotating disc having a slit. The slit transforms a continuous particle beam into an impulse. Different particles in the impulse with different velocities have also different times of flight from the chopper to the counter. A low signal-to-noise ratio is typical for this kind of measurement. This may be due to several causes, the main one being the low transmission of the chopper used in conventional time-of-flight experiments. To increase the signal-to-noise ratio a cross-correlation chopper has been developed. The chopper disc with one slit is replaced by a chopper having a set of slits arranged according to a PRBS. Cross-correlation modulators have been used in time-of-flight measurements in neutron scattering spectroscopy [12,13] and in particle beam experiments [14–18].

Using the multiple input in time-of-flight measurements enables us to obtain a better resolution of the velocity distribution of the particles. However, the method is extremely sensitive to chopper performance. Three severe requirements have hampered an extensive use of this technique: (1) exact reproduction of the theoretical PRBS on the chopper disc, (2) constant rotation period of the chopper disc and

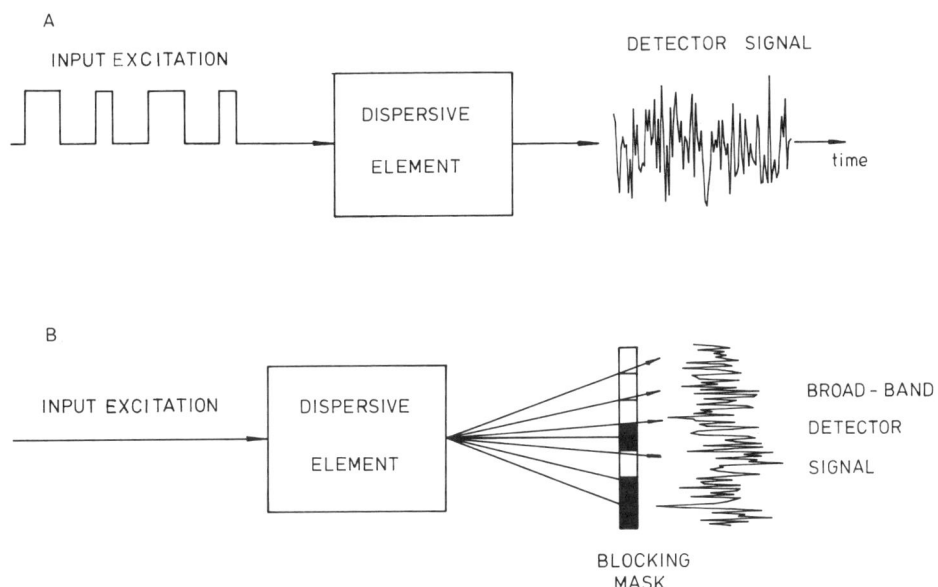

Fig. 2.2 — Multiplexing in time (A) and space (B).

(3) smooth rotation [17]. The poor fulfilment of these requirements leads to 'ghost' peaks and a noisy baseline. These difficulties have partly been overcome [17] by using a magnetically suspended cross-correlation chopper instead of a mechanically rotated chopper. The slits in the chopper were fabricated by using photoetching techniques instead of machining. In these experiments the input sequence theoretically consists of elements with values of 0 and 1. Due to the imperfections in chopper slit fabrication, the element values deviate from the theoretical values. In [17] the real values of the input elements were measured by using a laser beam and these were used in the calculations. Thus, at the price of extensive computations, the quality of the cross-correlation time-of-flight spectra was improved still further.

In time-of-flight measurements the impulse response function of the system is the particle velocity distribution. This is quite a similar experiment to chromatography where in fact the particles are separated according to their velocity through a medium. Ion mobility spectrometry is another example of this kind of measurement. However, we do not know of any reports of the use of PRBS in that field.

There are three forms of multiplex NMR: PRBS [19,20], stochastic [21,22], and fast scan or correlation [23,24]. In the first two forms the input is in the form of PRBS or a purely random sequence of radiofrequency pulses. The impulse response is found by the fast Hadamard-transform or by decorrelating the output from the input. This impulse response is equal to that obtained in pulse NMR and it requires an FT to get a conventional (continuous wave) NMR spectrum. The aim of multiplex NMR methods is to shorten the measurement time and to increase the signal-to-noise ratio. The sensitivity of the Hadamard-transform NMR method was found to be equivalent to that of FT–NMR but the advantage is its increased resolution because the duration of an excitation pulse determines not only the signal-to-noise ratio (S/N) but also the resolution: if the signal level is low, the excitation pulse duration in FT–NMR should be increased and that, in turn, decreases the resolution. In similar circumstances it is not necessary to increase the pulse duration in HT–NMR: if we assume the same S/N ratio in both FT– and HT–NMR, the resolution in the latter method is determined by the PRBS code length. We encounter the same situation when comparing correlation chromatography and conventional chromatography in which the signal amplitude is determined by the time during which the sample is introduced onto the column. Fourier-transform ion cyclotron resonance mass spectrometry with pseudo-random noise excitation was demonstrated in [25,26].

Correlation NMR is a method for the rapid gathering of data by a fast frequency- or field-sweep through the spectral range of interest. The scan time is usually only a few seconds (rapid scan NMR) and a conventional NMR spectrum is obtained by cross-correlating the rapid scan spectrum with the spectrum obtained by scanning the frequency through a single line under the same sweep conditions.

The fluorescence and photochemistry of a sample can be studied by use of a flash-lamp. If the sample is driven from equilibrium by a powerful light flash the resulting output will be masked by noise. A random input sequence of excitation light pulses can be exploited in these measurements to reduce pulse intensity and enable near-equilibrium operation of the system and determination of the rate of reversible photolytic reactions. In these experiments, a continuous laser beam is modulated by an electro-optical modulator to obtain the desired modulation waveform. A stochastic input was used by Haugen *et al.* for measuring the rate of photosensitive reactions [27]. The output function (kinetic decay curve) was obtained by using cross-correlation hardware. In other work, a PRBS input was used to study the decay of iodine fluorescence [28].

If an object does not luminesce or is not degraded photochemically, the absorbed energy is transformed into a heat pulse and gives rise to a heat flow through the sample material. In this way, thermal and structural properties of the sample can be studied. The method is kown as impulse response photoacoustic spectroscopy (also thermal wave imaging) and has been used for depth-resolved spectroscopic studies of solid samples [29–31]. To generate thermal input pulses, a PRBS-modulated laser beam was used. For good time-resolution and high spectral-depth resolution, a light pulse of short duration must be employed [30]. Under these conditions the quantity of heat generated in the sample is reduced, leading to a decrease in S/N ratio. Increasing the incident power increases the S/N ratio, but the use of this approach is limited by the maximum power of the laser and/or by damage to the sample. Thus, multiplex methods have good prospects in photoacoustic depth-resolved spectroscopy. Depth-resolved spectra of an artificially fabricated layer [30] and of dyed polymer films [31] have been studied by thermal PRBS-modulated excitation of the sample, and by decorrelation of the output.

In the examples above, the output function of a system was registered as a function of time. Spatial multiplexing (Fig. 2.2b) is also possible and widely used. This form of multiplexing the output signal requires some kind of a mask in front of the detecting device to block out certained spatial directions.

If a beam of particles is regarded as an input exciting variable, we have

Hadamard-transform beam and mass spectroscopy developed by Arikawa and co-workers [32–35]. The experimental arrangement was as follows: incident molecular ions (H_2^+ or H_3^+) were accelerated and introduced into a collision-cell containing target gases. The collision products, i.e. H_3^+, H_2^+, H^+ and H were assumed to have equal velocities, so mass-separation of the products was accomplished by a simple parallel-plate electrostatic analyser with an asymmetric geometry. The beam-encoding mask was located downstream from the analyser plate. The detector was located behind the mask. The mask had slits as in the case of time-of-flight measurements, which enabled the ions with a certain angle of deflection after the mass-analysing system to pass through the mask, and to be detected. A mask having a length of two PRBS should be used. During the experiment the mask position was shifted by means of a step motor by one slit width for each spectrum point measurement. So, since approximately 50% of the ions pass through the mask, detection efficiency increases. Computation of a conventional mass-spectrum is similar to that described for the weighing experiment.

Photoacoustic imaging of a surface involves scanning the sample with a focused laser beam and recording the signal for one picture element at a time. The energy needed to image one picture element, typically of the order of 1 mJ, limits the use of the method for light-sensitive samples, because of either light-induced degradation or too long an exposure time [36]. However, many spatial elements of the sample can be illuminated together through a mask, thus increasing the overall photoacoustic signal and at the same time keeping the local density of the light energy over one picture element low. This is the idea of Hadamard-transform photoacoustic imaging. The usefulness of the multiplex measurement method using Hadamard masks with 15 elements was described in [36]. The method enables characterization of a sample surface which has different absorbing regions. This method should also be useful in thin-layer chromatography.

In particle beam and photoacoustic HT spectroscopy, only preliminary experiments were performed, with masks of very low resolution. Mask fabrication is not an easy task and the aim of the investigations was to demonstrate the possibilities of multiplex measurements in these particular fields. To make full use of the advantages of multiplex methods in HT mass spectroscopy and photoacoustic imaging, mask resolution should be considerably increased, i.e. a PRBS with a large number of elements should be used. Further progress has been achieved in fabrication of the masks in HT optical spectroscopy, where masks with 2048 elements are in use today.

HT optical spectroscopy was invented in the late 1960s [37,38], but optical multiplexing had already been used in the late 1940s by Golay [39], who become known as the inventor of open tubular capillary columns in GC. HT optical spectroscopy has been described in [40] and several reviews, for example [11] in Chapter 7. HT spectrometers are available and are basically dispersive spectrometers with an encoding mask at the entrance and/or exit slit. The optical analyser is a prism or grating that spatially separates the incident radiation into its optical components. The opaque mask has slits or transparent regions which allow transmission of the dispersed light. The transmitted light is refocused on to the detector. The mask slits are cut according to the weighing experiment matrix already described. HT spectrometry requires as many measurements to be made as there are slits. On the other hand, the number of slits determines the resolution in the spectra.

Hence, as mentioned above, masks with a large number of slits are used. Conceptionally, HT optical spectrometry is equivalent to HT mass spectrometry.

The mask motion in HT spectrometers needs to be less carefully controlled than the mirror drive in an interferometer, but these spectrometers have lower resolution than interferometers. The instruments can be made portable and may find wide use in relatively low-cost applications where only medium resolution is required. HT spectrometers are used in the infrared domain where the detector noise is additive. In the UV–visible spectral region, the prime noise source is photon noise, which makes multiplexing ineffective (source noise). Also, the computer software is less complicated in HT spectroscopy than in FT spectroscopy, and because of the fast HT algorithm the spectra can be calculated very quickly.

In surface spectroscopy, a spectrometric imager has been developed. In one form of this device, a two-dimensional encoding mask is situated in the focal plane of a camera telescope and spatially encodes the image seen. The radiation then passes through the spectrometer, where it is coded again. The Hadamard-transform imaging spectrometers show promise for collecting spectral data about the earth's surface from a spacecraft. These devices have been developed for optical spectroscopy [41]. The HT X-ray telescope consists of a Hadamard mask and position-sensitive detector [42,43]. The device is useful for X-ray imaging of the sky, but is applicable also to plasma diagnosis by emission and to medical diagnosis by gamma rays.

Recently a new Hadamard-transform long-path absorption spectrometer was proposed in which an array of light-emitting diodes is used as the light source and a photodiode array is used as the detector [44]. This device is useful for high-accuracy measurements of atmospheric trace species. Another recent application of the multiplex measurement method was in Hadamard-transform photothermal deflection spectroscopy [45], a popular area of investigations in the last few years.

The variety of multiplex techniques can be classified according to the nature of the input excitation and the nature of the argument of the output distribution function. Table 2.1 presents a classification of the methods described in Section 2.1.2.

This list of applications is, of course, not complete and various other multiplex methods can be found [1,10,11].

As seen from this section, the multiplex advantage can be exploited in several ways, depending on the aim of the experiment:

(1) S/N ratio enhancement,
(2) lowered input excitation power level (with maintained S/N ratio),
(3) increased time, spectral, and spatial resolution (with maintained S/N ratio),
(4) decreased analysis time (as compared to ensemble averaging).

Feature (1) is essential and features (2) and (3) are in fact manifestations of the first. Feature (2) requires some comments. Although claimed in some places in the present chapter (by reference to the corresponding authors), it is not self-evident and requires further investigation. Because the mean input function level in the case of the PRBS input is far higher than the corresponding single-input excitation ampli-

Theory of correlation chromatography

Table 2.1 — Classification of multiplex mathods

Input (excitation) variables	Argument of output function	
	temporal	spatial or spectral
particles	chromatography time-of-flight spectroscopy	HT mass spectroscopy
electromagnetic radiation	HT–NMR stochastic-NMR correlation-NMR optical correlation spectroscopy ion cyclotron resonance with PRBS input	HT–infrared imaging HT–optical spectroscopy HT-X-ray imaging HT–absorption spectroscopy
thermal pulse	thermal wave imaging	HT–photoacoustic spectroscopy

tude (of which PRBS is composed), the system can be driven to the non-linear response region and complications arise immediately.

2.2 THEORY OF CORRELATION CHROMATOGRAPHY

2.2.1 Input–output relationship and a circulant matrix

In the previous section, we saw that in the case of multitiplex methods, the output is obtained by multiplexing the impulse response sequence by some kind of transformation matrix. The form of this matrix can be obtained from Eq. (1.2) by digitizing the corresponding functions $y(t)$, $x(t)$ and $h(t)$. Here, however, we prefer a longer but maybe clearer derivation of the input–output relationship in the case of CC (from our own point of view).

Let us assume that the input function $x(t)$, the chromatogram $h(t)$ and the output signal $y(t)$ are digitized with an interval Δt, the duration of which should be chosen according to the sampling theorem (see Section 3.2.1). Let us assume that the chromatogram $h(t)$ has n resolution elements measured, and so has $x(t)$: $x(t)$ is regarded as a constant within the interval Δt. Thus we have to deal with the following sequences (or vectors):

input: $\quad X = \{X_1, X_2, X_3, \ldots, X_n\}$

chromatogram: $H = \{H_1, H_2, H_3, \ldots, H_n\}$

output: $\quad Y = \{Y_1, Y_2, Y_3, \ldots, Y_n, \ldots\}$

With PRBS as the input pattern, an input sequence element, X_i, is generated according to the corresponding PRBS element; $X_i = 1$ means a sample is injected onto the column and $X_i = 0$ means the carrier gas is injected into the column at the time $i\Delta t$.

Let us assume that two input sequences are performed in series. The first injection, X_1, makes its contribution to the dectector output at the beginning of its first element, Y_1. Although we cannot see any peaks before elapse of the gas hold-up time of the column, the contribution of the first injection to the first detector output sequence element is frequently the pressure-disturbance peak amplitude. The speed of travel of the pressure disturbance through the column may be considered infinitely fast because of the relatively slowly moving concentration pulses. The second injection, X_2, makes its contribution to the output starting with the second element, Y_2 etc. The nth injection begins to contribute at the output, starting with the nth element of the output sequence. The next input sequence begins to contribute from the $(n+1)$th element of the output signal. The output sequence elements can now be expressed as follows:

$$\begin{aligned}
Y_1 &= X_1 H_1 \\
Y_2 &= X_1 H_2 + X_2 H_1 \\
Y_3 &= X_1 H_3 + X_2 H_2 + X_3 H_1 \\
&\overline{\qquad\qquad\qquad\qquad\qquad} \\
Y_n &= X_1 H_n + X_2 H_{n-1} + X_3 H_{n-2} + \ldots + X_n H_1 \\
Y_{n+1} &= X_1 H_1 + X_2 H_n + X_3 H_{n-1} + \ldots + X_n H_2
\end{aligned}$$
(2.3)

$$\begin{aligned}
Y_{2n} &= X_1 H_n + X_2 H_{n-1} + X_3 H_{n-2} + \ldots + X_n H_1 \\
Y_{2n+1} &= \qquad\quad X_2 H_n + X_3 H_{n-1} + \ldots + X_n H_2 \\
&\overline{\qquad\qquad\qquad\qquad\qquad} \\
Y_{3n-1} &= \qquad\qquad\qquad\qquad\qquad\qquad X_n H_n
\end{aligned}$$

The chromatogram vector H can be calculated by using the output elements between Y_{n+1} and Y_{2n} because the corresponding system of n linear equations has n unknowns (chromatogram vector elements). The first output elements, $Y_1, \ldots, Y_n$, cannot be used and the last $n-1$ values of Y_i are of no interest. Also, the input–output relationship can be derived by using only one input sequence. It is easy to see that in this case, the number of equations is $2n - 1$. By adding the corresponding equations for Y_i and Y_{i+n} the system of equations (2.3) can be 'folded back' to the system of n equations that coincides with the result we obtained for elements $Y_{n+1}, \ldots, Y_{2n}$ when using two input sequences. Further we shall deal with equations for output sequence elements $Y_{n+1}, \ldots, Y_{2n}$, using two PRBS inputs. If sequences $H = \{H_1, H_2, H_3, \ldots, H_n\}$ and $Y = \{Y_1, Y_2, Y_3, \ldots, Y_n\}$ are considered to be vectors, and the elements of the vector H in each equation are rearranged so that they are ordered from 1 to n, we can write the following matrix equation:

$$\begin{bmatrix} Y_{n+1} \\ Y_{n+2} \\ \ldots \\ Y_{2n} \end{bmatrix} = \begin{bmatrix} X_1 X_n & \ldots & X_2 \\ X_2 X_1 & \ldots & X_{n-1} \\ \ldots & \ldots & \ldots \\ X_n X_{n-1} & \ldots & X_1 \end{bmatrix} \begin{bmatrix} H_1 \\ H_2 \\ \ldots \\ H_n \end{bmatrix}$$
(2.4)

or $Y = XH$. Here X is a matrix formed from cyclically permuted input sequence elements.

The matrix X is known as a circulant matrix [46]. This matrix is determined by the first row (or column). Equation (2.4) is a digital representation of the convolution integral [Eq. (1.2)]. In its most general form, the vector of the chromatogram, H, can easily be found. The solution is $H = X^{-1}Y$, where X^{-1} is the inverse of X.

In fact, all multiplex methods can be expressed by Eq. (2.4). The matrix structure is different for different input functions. If we are dealing with a conventional, non-multiplex, method, then $Y = IH$ where I is an identity matrix and $Y = H$, i.e. we record a conventional chromatogram. On the other hand, if the input sequence is periodic, then the input sequence should have the following form: $X = \{1,0,1,0,1,0,1,0,\ldots,1,0\}$ or $X = \{1,1,0,0,1,1,0,0,\ldots,1,1,0,0\}$. The circulant matrices formed from the periodic type of input sequences have no inverse. These sequences cannot be used in multiplex experiments in which one detector is involved because $n - 1$ detectors are necessary for an n-component mixture (Section 1.2.2). It has been theoretically proved that for noise suppression, the best matrix for input is the Hadamard matrix [47]. One form of the Hadamard matrix has already been presented in describing the weighing experiment. With a PRBS as input sequence, the resulting matrix can be converted into a matrix that is closely related to the Hadamard matrix (see Appendix 2).

PRBSs have several properties making them very attractive. The PRBS is a sequence of ones and minus ones and the following autocorrelation property is important:

$$\sum_{j=1}^{n} X_{i \oplus j} X_j = \begin{cases} n & i \oplus j = j \\ -1 & i \oplus j \neq j \end{cases} \quad i = 0,1,\ldots,n-1 \tag{2.5}$$

The symbol $\oplus$ has the meaning of addition by the modulo, n, i.e. if $i + j > n$, then $i \oplus j = i + j - n$. Equation (2.5) seems to be complicated but it has a simple meaning: summing the sequence element squares equals n; summing the products of the corresponding elements of original and cyclically shifted (permuted) sequence elements results in -1. The PRBS is the best approximation of the random sequence that can be composed of a finite number of elements, because, as follows from statistics, a truly random sequence must have an infinite length. Sequences with the property of Eq. (2.5) can be generated by a special non-trivial algorithm considered in Appendix 2. As already seen in Section 1.3, PRBSs can be generated only with $n = 2^q - 1$ elements, where q is an integer. A PRBS contains $(n + 1)/2$ elements with values of $+1$ and $(n - 1)/2$ elements with a value of -1.

Let U be a circulant matrix composed of PRBS elements and J be a matrix with elements all equal to 1. For $n = 3$, U and J are as follows:

$$U = \begin{bmatrix} 1 & 1 & -1 \\ -1 & 1 & 1 \\ 1 & -1 & 1 \end{bmatrix} \quad J = \begin{bmatrix} 1 & 1 & 1 \\ 1 & 1 & 1 \\ 1 & 1 & 1 \end{bmatrix}$$

The following matrix equations are valid:

(1) $\quad UJ = JU = JU^T = U^TJ = J$

(2) $\quad J^2 = nJ$ \hfill (2.6)

(3) $U^T U = (n+1)I - J$

The first property can be proved by direct computation. Recalling the definition of the matrix multiplication and the fact that in PRBS the number of +1 elements is one more than that of the −1 elements, we obtain each element of UJ (and so on for the other products) by summing all PRBS elements. The sum is 1 for each element.

The proof of the second property can also be obtained by direct computation. An element in J^2 is obtained by summing n elements of J.

The third property follows from Eq. (2.5). The diagonal elements of the product $U^T U$ are obtained by summing the products of the corresponding elements of the non-permuted PRBS (i.e. by summing the squares of PRBS elements). The result is n. The non-diagonal elements are obtained by summing the products of the elements of permuted sequences. The result is −1. Recalling the definitions of the addition and multiplication of a matrix by a constant, we obtain the third property. If $n = 3$, we obtain:

$$U^T U = \begin{bmatrix} 1 & -1 & 1 \\ 1 & 1 & -1 \\ -1 & 1 & 1 \end{bmatrix} \begin{bmatrix} 1 & 1 & -1 \\ -1 & 1 & 1 \\ 1 & -1 & 1 \end{bmatrix} = \begin{bmatrix} 3 & -1 & -1 \\ -1 & 3 & -1 \\ -1 & -1 & 3 \end{bmatrix}$$

$$= 4\begin{bmatrix} 1 & 0 & 0 \\ 0 & 1 & 0 \\ 0 & 0 & 1 \end{bmatrix} - \begin{bmatrix} 1 & 1 & 1 \\ 1 & 1 & 1 \\ 1 & 1 & 1 \end{bmatrix}$$

If the input to the chromatograph is made according to a PRBS, then the −1 elements correspond to the injection of the pure carrier gas because injections with negative concentrations are not possible. The input sequence consists of the elements with values of 0 and 1 instead of −1 and 1. However, it is easy to see that in this case, Eq. (2.4) can be expressed as

$$Y = XH = \tfrac{1}{2}(U + J)H \qquad (2.7)$$

2.2.2 Decorrelation of the output sequence

Let us consider the method of inverting the input matrix, X, and computation of the chromatogram H. For convenience we call it decorrelation and the computed chromatogram a correlogram. This proposal is justified by the fact that the equation $H = X^{-1}Y$ means in fact a digital correlation of sequences X and Y as will be seen below. By analogy with the other multiplex methods, however, CC can also be called transform (Hadamard) chromatography.

As follows from Eq. (2.4) the input–output relationship in CC is in the form of a matrix equation with a circulant matrix. The circulant matrix can be diagonalized by the Fourier transform [46]. It means that $D = FXF^{-1}$, where F is the Fourier transform matrix and D is a diagonal matrix with non-zero elements along the main diagonal. The elements of the main diagonal are those of the Fourier transform of the input sequence. Taking into account the fact that $FF^{-1} = I$ we obtain from Eq. (2.4)

$$FY = FX(F^{-1}F)H$$

and

$$H = F^{-1}D^{-1}FY \qquad (2.8)$$

This is the matrix formulation of decorrelation by the Fourier transform described in section 1.4. Thus, decorrelation can be done by using a Fourier transformation. First, the output sequence is transformed, the result is divided by the elements of the Fourier transform of the input sequence, and at last, a correlogram is obtained by the inverse Fourier transformation of the result of division. With the fast Fourier-transform, decorrelation can be realized relatively easily.

In Eq. (2.8) the input function can be any function. The only necessary condition is that the FT of the input sequence must not contain elements equal to zero. For instance, in the case of the sinusoidal input function its Fourier transform contains only one non-zero element, corresponding to the sine function frequency and, as already mentioned, computation of the chromatogram is not possible.

If the input is in the form of a PRBS, then decorrelation can be directly done by multiplying Eq. (2.7) by the matrix U^T, i.e.

$$H = \frac{2}{n+1}U^TY = \frac{U^T}{n+1}(U+J)H = \frac{1}{n+1}[(n+1)I - J + J]H \qquad (2.9)$$

Thus, $X^{-1} = 2U^T/(n+1)$ in the case of the PRBS input with elements equal to 0 and 1. This is a rare case in linear algebra, when the matrix inversion has an analytical expression in a closed form. This is due to the unique properties of the PRBS. However, both decorrelation methods are in use in CC although no data are available to show which of them is better, either by Eq. (2.8) or by direct computation by Eq. (2.9). The fast Fourier-transform generally needs $n \log_2 n$ operations to be done, so for decorrelating by Eq. (2.8) we need $3n \log_2 n$ operations. Equation (2.9) requires n^2 operations for decorrelation but these operations are only additions and subtractions. At the same time, the fast Fourier-transform uses trigonometric functions extensively. The FT is a transform of complex numbers, which increases the number of necessary computations still more, although algorithms are known that take advantage of the fact that the input is a sequence of real numbers. Equation (2.8) can be used for any form of the input sequence, but Eq. (2.9) requires the use of a PRBS. It has been proved, however, that a PRBS input is best in reducing the detector noise [47]. Thus, the other input sequence forms, for example purely random noise, are not of much interest in CC. The straightforward implementation of Eq. (2.9) seems the easiest method of decorrelation. The situation can be still further improved by transforming Eq. (2.9) into the equation that involves the Hadamard matrix and taking advantage of the fast Hadamard-transform (FHT).

The FHT is a procedure for fast multiplication of the Hadamard type matrix by a vector. It has been developed for the PRBS input [19,47–50] and extensively uses the results of linear algebra and coding theory. The algorithm is quite complicated and its operation is explained in Appendix 2. Also in Appendix 2 the reader can find a description of feedback shift registers and the algorithm for generating PRBSs. All three topics are closely related. Appendix 3 gives a complete description of the related software for implementing the FHT algorithm.

2.2.3 Signal-to-noise ratio in correlation chromatography, and the multiplex advantage

The output signal of a detector contains noise. The noise can be any distortion of the output signal: high-frequency noise, drift, spikes, or even an unwanted signal from some sample component. Let us assume that this noise is additive, and independent of the output signal level. This is an important assumption. For some detectors noise depends on the signal level. This noise is called multiplicative. The following discussion is valid only for the additive noise.

Let us assume that in the presence of detector noise Eq. (2.4) can be written as

$$Y + R = X(H + \Delta H) \qquad (2.10)$$

Thus, the output signal is composed of the deterministic part Y and the random noise vector R. Decorrelation by Eq. (2.9) gives:

$$H + \Delta H = \frac{2}{n+1} U^T Y + \frac{2}{n+1} U^T R \qquad (2.11)$$

which is also composed of two parts: a 'pure' chromatogram and a noise part. Depending on the form of the disturbance, R_i, the noise contribution may be different. Generally, this contribution is in the form of a random baseline fluctuation and no analytical expression can be derived for it. Thus the mean and standard deviation are useful for the estimation of the disturbance level of a correlogram. From Eq. (2.1) we get for the correlation noise vector

$$\Delta H = \frac{2}{n+1} U^T R \qquad (2.12)$$

By the definition of the mean, we get from Eq. (2.12)

$$\overline{\Delta H} = \frac{1}{n} \sum_i \Delta H_i = \frac{1}{n} \left(\frac{2}{n+1} \right) \sum_{ij} U_{ij}^T R_j$$

Summing first over i and recalling that $\sum_i U_{ij} = 1$ we get

$$\overline{\Delta H} = \frac{2}{n+1}\overline{R} \tag{2.13}$$

Here $\overline{\Delta H}$ and $\overline{R}$ are mean values of the elements of the vectors ΔH and R respectively. Thus, the mean level of disturbance is reduced on the correlogram as compared to the disturbance level on the output signal.

For the variance of the ΔH elements we obtain by definition that

$$s_{\Delta H}^2 = \frac{1}{n-1}\sum_{i=1}^{n}(\Delta H_i - \overline{\Delta H})^2 \tag{2.14}$$

The sum of the squares of the ΔH elements is equal to the dot product $\Sigma \Delta H_i^2 = \Delta H^T \cdot \Delta H$ and for the variance $s_{\Delta H}^2$ we obtain

$$s_{\Delta H}^2 = \frac{1}{n-1}[\Delta H^T \cdot \Delta H - n(\overline{\Delta H})^2] \tag{2.14'}$$

Replacing $\Delta H^T \cdot \Delta H$ in Eq. (2.14′) by its expression in Eq. (2.12), and using Eq. (2.6), we obtain:

$$s_{\Delta H}^2 = \frac{1}{n-1}\left[\left(\frac{2}{n+1}\right)^2 R^T U U^T R - n\left(\frac{2}{n+1}\right)^2 \overline{R}^2\right]$$

Replacing $\overline{\Delta H}$ from Eq. (2.13) and because $R^T J R = n^2(\overline{R})^2$ we obtain

$$s_{\Delta H}^2 = \frac{1}{(n-1)}\left(\frac{2}{n+1}\right)^2[(n+1)R^T(\overline{R})^2 - n(n+1)(\overline{R})^2]$$

$$= \left(\frac{4}{n+1}\right)\left(\frac{1}{n-1}\right)[\mathbf{R}^T\mathbf{R} - n(\overline{R})^2]$$

As follows from Eq. (2.14), the expression in brackets is the variance of the detector output disturbance, s_R^2. Finally, we obtain for the standard deviation of the $\Delta \mathbf{H}$ elements:

$$s_{\Delta H} = \frac{2}{\sqrt{(n+1)}} s_R \qquad (2.15)$$

Thus, the detector output disturbance signal is suppressed by a factor of $2/\sqrt{(n+1)}$ on the correlogram. Let us now consider a few forms of the output disturbance.

According to Eq. (2.15), the high-frequency noise standard deviation is less by a factor of $2/\sqrt{(n+1)}$ in the correlogram than in the chromatogram. Let us assume that a correlation experiment and a single injection to the chromatograph are made under the same conditions, i.e. the amounts of sample injected in the single injection and by one injection in a correlation element are equal. In this case, the chromatogram peak height is the same on the correlogram and on the single-injection chromatogram but since the noise level in the case of CC is lower than in the single-injection chromatogram, the S/N ratio in the correlogram is higher. This is the multiplex advantage of correlation chromatography over conventional chromatography and utilizing the multiplex advantage is the purpose of a correlation experiment. It is easy to calculate that to obtain an increase of about one order in S/N a 511-element PRBS should be used, and (recalling the derivation of the input–output matrix) 256 injections of the sample would be required. To increase S/N by a factor of 100, a 32768-element PRBS should be used and 16384 injections would be necessary. It follows that CC presents sampling requirements not previously encountered in chromatography. Currently an increase of S/N by one order is common in CC but there are some examples of increasing S/N by two orders. One demonstration of the multiplex advantage is given in Fig. 2.3 [51].

Spikes on the output are due to short duration electrical bursts caused mainly by an on/off switching of high-power electrical machines. These bursts travel along the earth wire and appear on the detector signal as a disturbance. A spike can be considered as a vector having all elements equal to zero except for one, equal, say, to z. The mean of the spike vector elements is z/n and the standard deviation is $z/\sqrt{n}$. It follows from Eq. (2.15) that in the case of a spike $s_{\Delta H} = 2z/n$ for large n. Thus, the spike power, as expressed in terms of variance, is reduced. However, the spike on the detector output is transformed into a baseline fluctuation on the correlogram. It can easily be calculated by Eq. (2.12) that the baseline noise on the correlogram has a PRBS pattern with a certain phase shift. A large spike can easily be recognized from the detector output and removed before decorrelation. An example of this procedure is shown in Fig. 2.4.

Baseline drift is a common phenomenon in chromatography. For a linear drift, the disturbance vector can be regarded as having elements $R_i = a + ib$. Here a is a

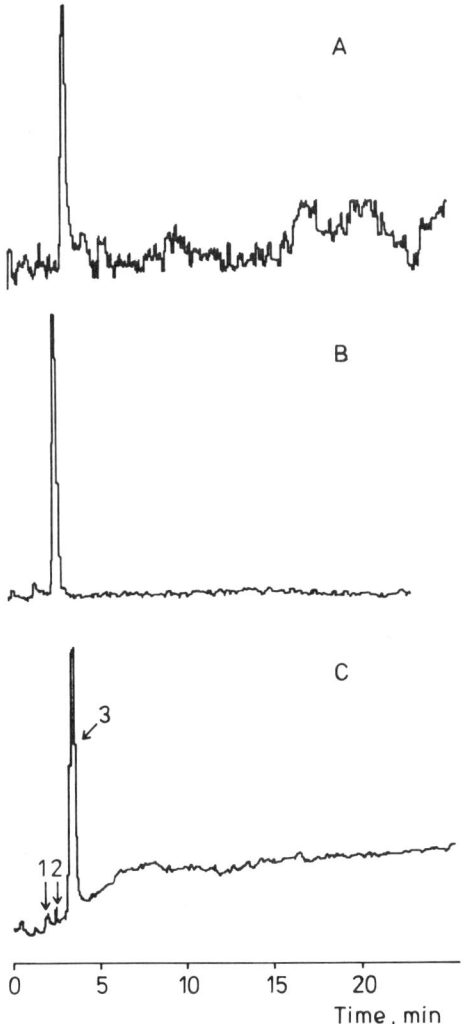

Fig. 2.3 — Correlogram (B) and single-injection chromatograms (A) and (C) of ethanol. (A) and (B), the same amounts of sample; (C), ten times as much as sample as for (A) and (B). During correlogram run (B) 1024 injections were made. Peaks: 1, unknown, 2, water; 3, ethanol. TCD and a column with selective polymer sorbent Anteparon 22 (Antechnica) were used. Column temperature 112°C; sample was introduced by Deans-type switch from the carrier flow at 0.1 ml/sec [51]. (Reprinted with permission from M. Kaljurand and E. Küllik, *Trends Anal. Chem.*, 1985. Copyright 1985 Elsevier Science Publishers.)

constant baseline offset and the drift speed is given by the factor b. The mean and standard deviation of the drift vector elements are

$$\bar{R} = a + \left(\frac{n+1}{2}\right)b$$

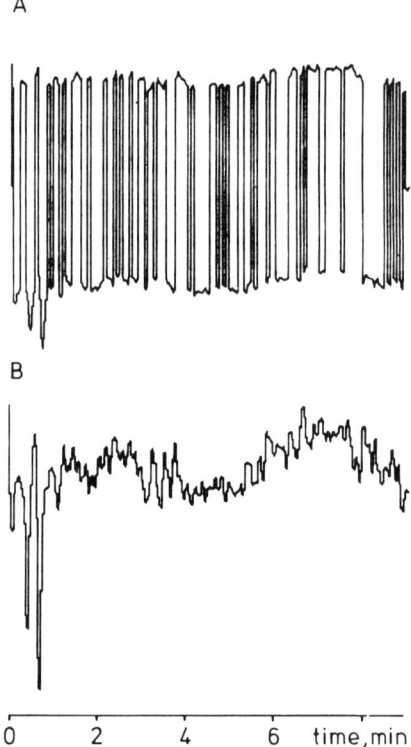

Fig. 2.4 — Effect of spike on correlogram. Correlogram (A) was computed from the detector output containing a single spike with an amplitude of 2043 units. The mean detector output level was 43 units. Correlogram (B) was computed after spike removal. Sample: aniline (first peak) and toluidine (second peak) in a steam flow with FID detector. Steam–solid chromatography.

$$s_R = b\sqrt{\frac{(n+1)n}{12}}$$

A linear drift in the detector output becomes a specific form of fluctuation of the baseline on the correlogram. The mean and standard deviation of this distortion are obtained from Eqs. (2.14) and (2.15) as follows:

$$\overline{\Delta H} = 2a/(n+1) + b$$

$$s_{\Delta H} = b\sqrt{n/3}$$

(2.16)

It follows from Eq. (2.16) that a constant baseline offset does not present problems in

Sec. 2.2] **Theory of correlation chromatography** 47

CC, although the drift makes a significant contribution to the correlogram noise standard deviation for large b or n values and thus is a dangerous phenomenon in CC. It is useful to subtract the baseline from the detector output before decorrelation. One example of the influence of drift on the correlogram is given in Fig. 2.5 [52]. It follows from Fig. 2.5 that on the correlogram baseline drift becomes a specific half-wave that could impede the detection of a useful signal. Subtracting the baseline of the detector output from the signal removes the problem.

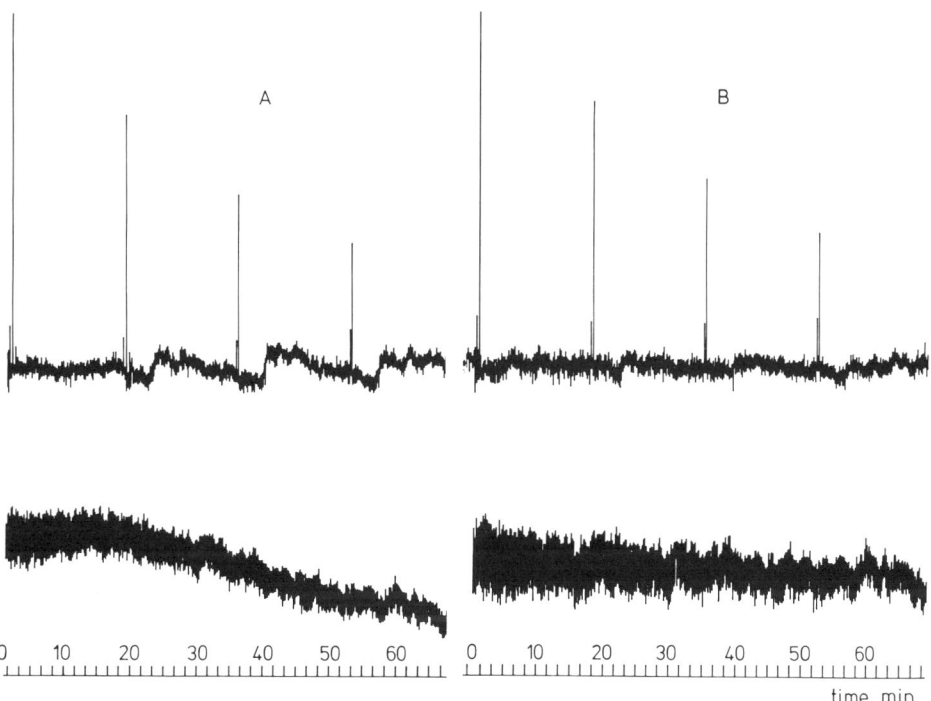

Fig. 2.5 — Effect of baseline drift on correlograms: (A) four consecutive methane correlograms (upper trace) and corresponding detector output (lower trace); (B) the same as (A) but with baseline drift removed. (Adapted from [52].)

Sinusoidal oscillations of the baseline can be generated by the power supply of the chromatograph. On a correlogram a sinusoidal oscillation amplitude is suppressed according to Eq. (2.15). It is surprising, however, that a sinusoid on the detector output remains a sinusoid on the correlogram baseline, with the same frequency but with a different phase and a suppressed amplitude. This was first shown by Sloane *et al.* [53].

As a continuous detector output voltage is converted into the sequence of numbers in the computer in CC by an analogue-to-digital converter (ADC), the quantization noise also influences correlograms. The ADC is characterized by the number of quantization levels and the voltage differences that can be measured by it (the 'least significant bit', LSB value). Let the LSB be denoted by q. If y_1 and y_2 are two detector outputs, then, if $|y_1 - y_2| < q$, the ADC does not differentiate between these two voltage values, i.e. we have an error or digitization noise. The mean of this noise is zero, the variance is $q^2/12$. The quantization noise is uniformly distributed in the interval $[-q/2, q/2]$ [54]. The quantization noise level is suppressed in the correlogram according to Eq. (2.15). This property can be advantageously used to lower the quality requirement for the ADC: if a certain problem requires an LSB value of, say, q volts, the same precision could be obtained by using an LSB of $\frac{1}{2}q\sqrt{n+1}$ volts, which is equivalent to an increase in the resolution of the ADC with a given number of levels as compared to single measurements. This increase is significant because the cost of an ADC depends strongly on the number of quantization levels and ADCs with a high number of levels are expensive. Similar ideas are used to increase the resolution of the ADC by using esemble averaging and adding random noise to the measured voltage [55]. The simplest ADC is a device with two levels — the signal polarity detector. Thus, the possibility of the correlation chromatograph with a very simple ADC appears. But, as follows from Belchamber and Horlick [55] this device only works properly in very low signal to noise situations.

Contributions of various output signal disturbances to the correlogram are summarized in Table 2.2.

Table 2.2 — Contribution of various output signals to disturbances on the correlogram

Baseline disturbance	Noise form in correlogram	Noise vector elements	Correlogram noise standard deviation†
high-frequency random noise	random noise	$\{R_i\}$	$2s_R/\sqrt{n}$
spikes	PRBS pattern	$\{0, \ldots, z, \ldots, 0\}$	$2z/n$
linear drift	half-wave	$\{a + bi\}$	$b\sqrt{n/3}$
sinusoidal oscillation	sinusoidal oscillation	$\sin(2\pi f i)$	$\sqrt{2/n}$

[53]

† Approximations for large n values.

2.2.4 Non-stationary input

Correlation chromatography assumes that the chromatogram function, $h(t)$ and the input sequence do not change during the multiple inputs. This is the stationary system assumption, which unfortunately is not fulfilled in real-life situations. Four types of non-stationary system are encountered in practice.

(1) Random variation of the amounts of sample injected. This type of non-stationary input is caused by the non-ideality of the sampling device and also by the random variations in the flow composition.
(2) Systematic variation in the amount of sample injected. This type of non-stationary input appears in calibration of the system by some dynamic method for generation of the sample. An exponential dilution flask generates a flow with

a sample concentration decreasing exponentially with time. A diffusion cell generates a flow with a concentration that decreases proportionally to $1/\sqrt{t}$ where t is the experimental time (see Section 4.2.1). In this type of input the chromatogram function, $h(t)$, remains unchanged during the experiment. The amplitudes of all peaks change in time according to the same function.

(3) Each input flow component changes in a different way. This case is common in studying reactions by CC.
(4) The variations in the flow composition are generated by the chromatographic system itself. Strictly speaking, this is not a non-stationary input, but from the point of view of automatic control theory, it can be considered as such. Variations in the flow are generated by variations in the carrier gas velocity and in the column temperature. Also variations in the changes in the mobile phase composition (e.g. in steam–solid chromatography) contribute to this kind of non-stationary effect.

If the correlogram is computed from a detector output recorded for a non-stationary input, there will be a particular type of disturbance in the chromatogram: the correlation noise. For the first two types of non-stationary input, the standard deviation of the correlation noise can be calculated, but the other cases are difficult to describe mathematically. Below, we consider all four non-stationary inputs in more detail.

2.2.4.1 *Random variations in the sampled quantities of the flow*

Let us assume that instead of the ideal input sequence $\{X_1, X_2, \ldots, X_n\}$ a real input sequence $\{X_1 + E_1, X_2 + E_2, \ldots, X_n + E_n\}$ is used in Eq. (2.3). Here E_i is a sampling error and if $X_i = 0$ then $E_i = 0$ also, but if $X_i = A$ where A is the amount of sample injected then E_i is some fraction of A. Hence we can assume that the mean value of the non-zero input sequence elements is A, with standard deviation s_E. Performing two PRBSs simultaneously as usual, we get the second input sequence with elements $\{X_1 + E_{n+1}, X_2 + E_{n+2}, \ldots, X_n + E_{2n}\}$. Though the input sequence is periodic with period n, the error of the sampling is not assumed to be periodic. As in Section 2.2.1 we can write a system of linear equations similar to Eq. (2.3), using the input sequences with error elements E_i. Instead of Eq. (2.4) we get the following matrix equation:

$$\mathbf{Y} = \mathbf{XH} + \begin{bmatrix} E_{n+1} & E_n & E_{n-1} & \cdots & E_2 \\ E_{n+2} & E_{n+1} & E_n & \cdots & E_3 \\ E_{n+3} & E_{n+2} & E_{n+1} & \cdots & E_4 \\ \cdots & \cdots & \cdots & & \cdots \\ E_{2n} & E_{2n-1} & E_{2n-2} & \cdots & E_{n+1} \end{bmatrix} \begin{bmatrix} H_1 \\ H_2 \\ H_3 \\ \cdots \\ H_n \end{bmatrix} \qquad (2.17)$$

or $\mathbf{Y} = \mathbf{XH} + \mathbf{EH}$. Matrix $\mathbf{E}$ is known as the Toepliz matrix [46], which does not have such good properties as the circulant matrix. For example, it cannot be inverted by a Fourier transform.

Decorrelating the output as usual by multiplying the output vector $\mathbf{Y}$ by the matrix $2\mathbf{U}^T/(n+1)$ we obtain:

$$\frac{2}{n+1}\mathbf{U}^\mathrm{T}\mathbf{Y} = \mathbf{H} + \frac{2}{n+1}\mathbf{U}^\mathrm{T}\mathbf{E}\mathbf{H}$$

and for the correlation noise vector:

$$\Delta\mathbf{H} = \frac{2}{n+1}\mathbf{U}^\mathrm{T}\mathbf{E}\mathbf{H} \qquad (2.18)$$

Because the mean of E_i is zero the mean of the correlation noise vector, $\overline{\Delta H}$, is also zero, as follows from Eq. (2.18). The variance of the correlation noise $s^2_{\Delta H}$, is obtained by using Eq. (2.6)

$$s^2_{\Delta H} = \frac{1}{n-1}\Delta\mathbf{H}^\mathrm{T}\Delta\mathbf{H}$$

$$= \left(\frac{1}{n-1}\right)\frac{4(n+1)}{(n+1)^2}\mathbf{H}^\mathrm{T}\mathbf{E}^\mathrm{T}\mathbf{E}\mathbf{H} - \left(\frac{1}{n-1}\right)\frac{4}{(n+1)^2}\mathbf{H}^\mathrm{T}\mathbf{E}^\mathrm{T}\mathbf{J}\mathbf{E}\mathbf{H}$$

The last term of this last equation is zero because $\overline{\Delta H} = 0$. Thus the variance is

$$s^2_{\Delta H} = \frac{4}{(n-1)(n+1)}\mathbf{H}^\mathrm{T}\mathbf{E}^\mathrm{T}\mathbf{E}\mathbf{H} \qquad (2.19)$$

The product $\mathbf{E}^\mathrm{T}\mathbf{E}$ can be approximated for large n as $\mathbf{E}^\mathrm{T}\mathbf{E} = ns^2_E\mathbf{I}/2$. This is because all E_i are assumed to be uncorrelated random variables and because of this, only main diagonal elements are dominant in $\mathbf{E}^\mathrm{T}\mathbf{E}$. The elements in the main diagonal have the value ΣE_i^2 and because the mean of E_i is zero all the main diagonal elements are approximately equal to $(n-1)s^2_E/2$. The factor $1/2$ appears because only half the E_i values are not equal to zero. Thus we get from Eq. (2.19) the expression

$$s^2_{\Delta H} = \frac{2}{n+1}s^2_E\mathbf{H}^\mathrm{T}\mathbf{H} \qquad (2.20)$$

Assuming that the chromatogram has only one Gaussian peak, with standard deviation σ, the product $\mathbf{H}^\mathrm{T}\mathbf{H}$ value can be estimated. $\mathbf{H}^\mathrm{T}\mathbf{H}$ is a dot product of the chromatogram sequence elements $\{H^1, H_2, \ldots, H_n\}$. Thus by definition we obtain:

$$\boldsymbol{H}^{\mathrm{T}}\boldsymbol{H} = \sum_{i=1}^{n} H_i^2 \leq \frac{1}{\Delta t}\sum_{-\infty}^{\infty}\Delta t\left\{\exp\left[-(i\Delta t)^2/2\sigma^2\right]\right\}^2$$

Here Δt is a digitization interval for the chromatogram. The value of the sum can be approximated by the integral. Assuming $\Delta t \to 0$, the value of the integral is $\sigma\sqrt{\pi}$.

Thus for the standard deviation of the correlation noise we finally obtain, from Eq. (2.20),

$$s_{\Delta H} = \sqrt{\left[\frac{2\sigma\sqrt{\pi}}{(n+1)\Delta t}\right]} s_E \tag{2.21}$$

If there are more peaks than one in the chromatogram then each peak generates its own correlation noise on the chromatogram and the total correlation noise variance can be found by addition of the individual peak variances.

It follows from Eq. (2.21) that the standard deviation of the correlation noise is determined by the input noise as well as by the peak width. The wider the peak, the larger the noise amplitude. Also the digitization interval is important. Oversampling of the peak is not wise. From the sampling theorem (Section 3.2.1), we find that $\Delta t = \sigma/2$ is a suitable value for the digitization interval for a Gaussian peak and in that case $s_{\Delta H} = 2\pi^{1/4} s_E/\sqrt{(n+1)}$. The input noise is suppressed as well as the detector high-frequency noise but to a lesser extent (approximately by a factor of $\sqrt{\sigma/\Delta t}$.

Correlation chromatography can be used not only in situations where detector noise is a limiting factor, but also where input noise reduction is important. Let us look at measurement of the response factors. To get precise values for these a lot of replicate measurements have to be made to increase the precision — a time-consuming procedure. Correlation chromatography offers far higher precision of the measurement in a time that is comparable to the recording of two chromatograms. Another example is the study of chemical reactions by correlation chromatography. The reaction rate is controlled by temperature, variations in which generate variations of the amount of products evolved. In thermal degradation reactions the temperature variations are critical in influencing the reaction rate. The CC suppresses these variations to a great extent and precise correlograms are recorded.

2.2.4.2 Systematic variations in the sampled quantities of the flow

The mathematical description of CC when the input is changing systematically according to some smooth function becomes very complicated, and the conclusions that can be derived are of little use for improving the quality of the correlogram i.e., for reducing the correlation noise. The problem is generally recognized as 'identification of process varying with time' and the common approach is to develop the input function as a Taylor series expansion [56,57]. In correlation chromatography this approach was applied in [57–60]. The expressions that describe the relationship between input and output are complicated and useless because their evolution requires knowledge of the Taylor series expansion coefficients, which are generally unknown. The computational complexity increases remarkably.

For the linearly changing input function $X_i = a + bi\Delta t$ and a single narrow chromatographic peak, the following approximate formula for the standard deviation of the correlation noise has been derived [52].

$$s_{\Delta H} = \gamma b\sqrt{n}\Delta t \tag{2.22}$$

where γ is a dimensionless coefficient that depends on the peak position and the peak width. γ varies between 0.5 and 1.5 and for the common chromatographic siutations it can be taken as equal to unity, i.e. $s_{\Delta H} \approx b\sqrt{n}\Delta t$. Hence, the total input variation over one input sequence, $b_{n\Delta t}$, is suppressed according to the $\sqrt{n}$ rule. However, the S/N ratio on the chromatogram is important and, according to Eq. (2.22), is $S/N = a/b\sqrt{n}\Delta t$. If a long input sequence is used the S/N ratio decreases. Systematic non-stationary states generate correlation noise that cannot be reduced. Baseline drift has a similar effect, and the correlation noise standard deviation increases proportionally to $\sqrt{n/3}$. The only parameter that can be used to reduce the correlation noise is the process rate, b, and by using Eq. (2.22) it is possible to estimate the process rate necessary to record a chromatogram with an acceptable signal to correlation noise ratio. This is important for the calibration of CC by some dynamic method, e.g. by exponential dilution of the sample.

The description of the non-stationary input in the case when each input flow component varies according to its own law (or the non-stationary state is caused by variations in the chromatographic system) is complicated. If the variations are systematic and can be described by some smooth function with all derivatives known, then we can expand the input function, $x(t)$, as a Taylor series. A change in the linear input function is the same for all chromatogram elements H_i and that enables use of a transform to describe the change of the input sequence. In that case each chromatogram sequence element must be considered differently. If we have some *a priori* information about the non-stationary behaviour of the system, the non-stationary conditions can be compensated to some extent, though at the cost of extensive computation. If the derivatives of the Taylor series expansion are known then it is possible to generate the corrections for the output vector elements. This approach was used in [57]. It has limited application, however, because the derivatives are usually not known. The exception is perhaps the calibration of the correlation chromatograph by a dynamic method with some known function of input concentration.

Another method of correction for non-stationary conditions has been developed and used in CC by Kaljurand and Küllik [58] who assumed that the chromatograph elements can be expanded as a Taylor series. A knowledge of the derivatives of the series is not necessary and the non-stationary conditions can be compensated to whatever precision is desirable but at the cost of computational and experimental complexity. In this approach several PRBS inputs are used. The non-stationary conditions of the system are estimated by making a polynomial fit to the corresponding elements of the output vector (spaced by multiples of one sequence). By doing this, an ouput matrix instead of the output vector can be constructed. Multiplying the output matrix by $2\mathbf{U}^T/(n+1)$ gives the non-stationary impulse response matrix.

As can easily be seen, the computational algorithm enables computation of a two-dimensional chromatogram if the behaviour of the chromatogram with time is suitable. It must not change very rapidly during one PRBS cycle. The performance of this algorithm was studied by using simulated chromatograms [58]. The practical use of this algorithm is restricted by the computational difficulties because it is necessary to compute n^2 elements of the two-dimensional chromatogram, which is a complicated and lengthy task, even for todays's personal computers.

Another peculiarity of this algorithm is that reduces the multiplex advantage of the CC method. It has been proved [56] that if we use the piecewise linear approximation of the output sequence elements the multiplex advantage (lowering of the detector noise) is reduced by a factor of 2.67 and for quadratic interpolation this factor is 6.8. We will return to this algorithm in Section 3.2.3 when dealing with correlation chromatography and non-stationary flow composition.

To estimate the real value of this algorithm, further studies are necessary. The algorithm seems to have a good potential for compensating for smooth variations in the chromatograph performance during a CC experiment. For measurements under non-stationary flow conditions, however, better methods exist and will be described in Chapter 3.

2.2.5 Validation of the formulae for correlation noise

The formulas for the correlation noise can be validated by comparing theoretical results with experimental ones. However, in some cases it is difficult to generate a distortion of known amplitude on the detector output. In that case CC should be simulated on a computer.

The validity of the general formulae for the multiplex advantage, Eq. (2.15), is experimentally confirmed by many investigators. We refer to Fig. 2.3 (p. 45) where single-injection chromatograms of a known amount of sample are compared with a correlogram. The theoretically expected improvement in S/N is a factor of $\sqrt{1024}/2 = 16$, and is completely fulfilled. The validity of the $\sqrt{n}/2$ noise suppression rule has also been confirmed by computer simulation experiments [50].

The generation of an experimentally known rate of baseline drift is not easy, so CC was simulated on the computer. The detector output was generated and a linear function was added to it. The composite output was decorrelated and the standard deviation of the correlation noise was computed by means of Eq. (2.14). The sequence length was varied from 128 to 1024 for two drift rate values. The correlation coefficient between calculated [Eq. (2.14)] and theoretical [Eq. (2.16)] values was 0.99995.

A similar procedure was applied for the validation of Eq. (2.21), the variables in which were varied between the limits $n = 127–1024$, $\sigma/\Delta t = 1–4$ and $s_E = 0.25–1.0$. The correlation coefficient between the theoretical and calculated values was 0.995, showing remarkably worse correlation than that in the baseline drift simulation. This is understandable because Eq. (2.21) is only an approximate formula. Simulation of the linearly changing input showed that Eq. (2.22) holds with better precision. The correlation coefficient between experimental and theoretical values is 0.9994.

Thus the correlation noise from the most important sources can be estimated with some precision and the success of the correlation experiment predicted.

2.3 COMPARISON OF CC WITH OTHER METHODS FOR THE MEASUREMENT OF LOW-CONCENTRATION SAMPLES

2.3.1 Injecting large samples

Consider a flow changed in composition by introduction of a sample during a time interval, Δt, onto the chromatographic column. The larger the interval Δt the larger the amount of substance injected onto the column and the larger the S/N ratio. However, increasing this interval to infinity leads not to an increase of the S/N ratio to infinity, but only to some constant value. On the other hand increasing Δt decreases the resolution of the chromatogram. These two effects limit the extent to which measurement of low concentrations of substances can be achieved by introducing large samples into the column. Let us consider this problem in more detail.

If a chromatographic peak has the Gaussian form with amplitude A and variance σ^2, the injection of the sample during the interval Δt results in a concentration distribution at the end of the column according to the function:

$$y(t) = \int_{-\Delta t/2}^{\Delta t/2} h(t-\tau)d\tau = \int_{t-\Delta t/2}^{t+\Delta t/2} h(\tau)d\tau \qquad (2.23)$$

Equation (2.23) follows from the convolution integral [Eq. (1.2)]. Here $y(t)$ is the detector output signal and $h(t)$ the chromatogram function for injection of an infinitely narrow concentration pulse. Function $y(t)$ has a maximum y_m, when $t = 0$ i.e.

$$y_m = A \int_{-\Delta t/2}^{\Delta t/2} \exp(-\tau^2/2\sigma^2)d\tau = \sqrt{2\pi}\sigma A \, \text{erf}\left(\frac{\Delta t}{2\sqrt{2}\sigma}\right) \qquad (2.24)$$

where erf (x) is the error integral [54]. The error integral can be approximated (with precision of 1%) for small argument values as erf $(x) = 2x/\sqrt{\pi}$ if $x < 0.176$. The last condition determines the linear region of this method of sample introduction: $\Delta t/\sigma \leq 0.5$. On the other hand $y_m \to \sqrt{2\pi}\sigma A$, if $\Delta t \to \infty$ and $y_m \leq 0.999 \sqrt{2\pi}\sigma A$ if $\Delta t \geq 6\sigma$. This condition determines the largest Δt value that still increases the signal-to-noise ratio. There is no gain in using larger values, because doing so will not significantly increase the signal-to-noise ratio.

On the other hand the Δt value is limited by the resolution desired. The peak total second moment σ_T^2 is determined by the additivity rule of the variances:

$$\sigma_T^2 = \sigma^2 + (\Delta t)^2/12$$

The second term $(\Delta t)^2/12$ is the increase in the chromatographic peak variance caused by the finite length of the injection time. If the increase of the peak variance is not very large then the relative increase of peak variance can be approximated as

$$\frac{\sigma_T^2 - \sigma^2}{\sigma^2} = \frac{2\Delta\sigma}{\sigma} = \frac{1}{12}\left(\frac{\Delta t}{d}\right)^2 \qquad (2.25)$$

where $\Delta\sigma = \sigma_T - \sigma$. The resolution, R, can be defined as $R = \Delta t_R/(2.355\sigma)$ where Δt_R is the difference in retention time between the two closest peaks in the chromatogram. The relative decrease in the resolution due to variation in σ is $\Delta R/R = \Delta\sigma/\sigma$, and by using Eq. (2.25) we can write the relative decrease in resolution as a function of the sampling time:

$$\frac{\Delta R}{R} = \frac{1}{24}\left(\frac{\Delta t}{\sigma}\right)^2 \qquad (2.26)$$

Equation (2.26) determines the longest Δt value that should be used to maintain an acceptable resolution. On the other hand, as follows from Eq. (2.24) this interval determines the amplitude of the peak and thus the signal-to-noise ratio in the chromatogram. These two requirements for Δt cannot be satisfied simultaneously.

To maintain the required resolution with increased signal-to-noise ratio CC can be used. This was pointed out already in Section 2.1.2. Usually in CC Δt is equal to the digitization interval and it follows from the sampling theorem (Section 3.2.1), that $\Delta t \leq 0.5\sigma$ is a good value for the injection time. The peak width increase is then 1% [as follows from Eq. (2.26)] and hence only gives a small decrease in the resolution. However, evaluating the numerical values for the error integral in Eq. (2.24) shows that the peak amplitude is only a fifth of the maximum attainable value for $\Delta t \to \infty$. Thus using a PRBS input (e.g. with $n = 127$ elements) gives a correlogram in which the resolution is not deteriorated but the signal-to-noise ratio is better then it is possible to obtain with a single injection of a large sample.

A practical example of the improvements in resolution is given in Fig. 2.6. Two chromatograms, A and C, have comparable signal-to-noise ratio. However, to obtain the required S/N ratio for a single-injection chromatogram (C) a very large Δt is required and consequently the peaks are much wider than in the case of CC (A) for the same sample. For comparison a single-injection chromatogram (B) with short Δt is given.

2.3.2 Sample concentration

It is often necessary to increase the concentration of a component. This can be realized either by collecting the compound in a cold trap or on a specific adsorbent, or by extracting with a suitable solvent. There are several drawbacks of the concentration method.

(1) Concentration by adsorption is a special problem; each sample class requires its own type of adsorbent. Sometimes this is, of course, not a drawback but an advantage when it is necessary to determine special target compounds and very specific adsorbents can be developed.
(2) When the sample contains reacting components, the reaction rates will increase

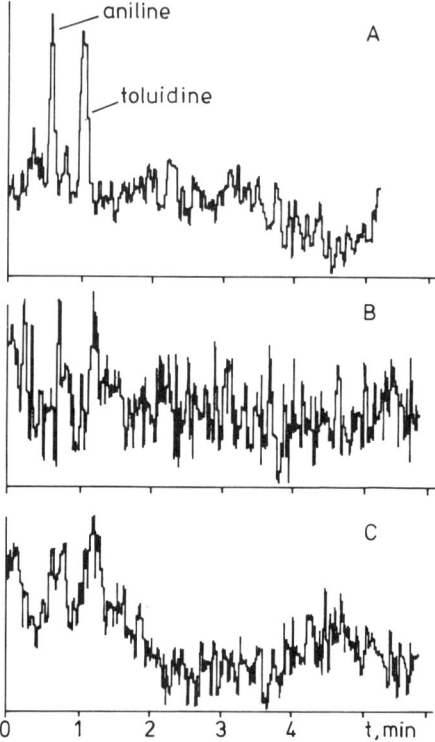

Fig. 2.6 — Multiplex advantage. Improved resolution of correlogram compared with single-injection chromatogram: (A) correlogram, sampling time 1.3 sec; (B) single-injection chromatogram, sampling time 2.9 sec; (C) single-injection correlogram, sampling time 11 sec; (steam–solid chromatography of aniline and toluidine sample).

if the concentration of the components increases, and thus the sample composition changes.
(3) Sample recovery from a trap is never complete and some of the sample remains in the trap.
(4) Micro fog-formation in a trap can cause loss by breakthrough.
(5) If the sample is a mixture of substances from different classes, then their adsorption properties will differ and there is no guarantee that all components will be retained to the same extent by a single adsorbent.

These features of the concentration method are not problems in correlation chromatography. Because it is a direct method the sample passes through the chromatographic system without any change except reaction in the detector. In principal no sample pretreatment is necessary. However, if advantage is taken of the properties of some adsorbents developed in chromatography, trace compounds can be determined in the presence of a large amount of matrix compounds by removing the latter by adsorbents before the injection.

Most media can be used as a carrier in correlation chromatography, e.g. air in determination of atmospheric impurities, or steam in determination of trace compounds in water.

Some applications of CC for solving environmental problems will be described in Section 5.2. Correlation chromatography seems to be well suited for environmental applications, because large amounts of low-concentration samples are available. Also the instruments can be relatively simple if advantage is taken of the sophisticated data-processing capabilities of modern personal computers.

REFERENCES

[1] L. M. Soroko, *Multiplex Measurement Systems in Physics*, Atomizdat, Moscow, 1980 (in Russian).
[2] N. J. Sloane and M. Harwit, *Appl. Opt.*, 1976, **15**, 107.
[3] F. Yates, *J. Roy. Stat. Soc., Suppl.*, 1935, **2**, 181.
[4] P. Fellgett, *J. Phys. Radiol.*, 1958, **19**, 187.
[5] H. Fizeau, *Ann. Chim. Phys.*, 1862, **66**, 429.
[6] A. A. Michelson, *Phil. Mag.*, 1892, **34**, 280.
[7] J. W. Cooley and J. W. Tukey, *Math. Comput.*, 1965, **19**, 297.
[8] R. J. Bell, *Introductory Fourier Transform Spectroscopy*, Academic Press, New York, 1972.
[9] T. C. Farrar and E. D. Becker, *Pulse and Fourier Transform NMR*, Academic Press, New York, 1971.
[10] P. R. Griffiths (ed.), *Transform Techniques in Chemistry*, Plenum Press, New York, 1978.
[11] A. G. Marshall (ed.), *Fourier, Hadamard and Hilbert Transforms in Chemistry*, Plenum Press, New York, 1982.
[12] K. Sköld, *Nucl. Instrum. Methods*, 1968, **63**, 114.
[13] W. Raith, *Advances in Atom and Molecular Physics*, Vol. 12, p. 281. Academic Press, New York, 1976.
[14] C. Nowikow and R. Grice, *J. Phys. E.*, 1979, **12**, 515.
[15] C. A. Visser, J. Wolleswinkel and J. Loss, *J. Phys. E.*, 1970, **3**, 483.
[16] G. Wilhelmi and F. Compf, *Nucl. Instrum. Methods*, 1970, **81**, 36.
[17] G. Comsa, R. David and B. J. Schumacher, *Rev. Sci. Instrum.*, 1981, **52**, 789.
[18] A. M. Ducorps and C. J. Yashinovitz, *Rev. Sci. Instrum.*, 1983, **54**, 444.
[19] R. Kaiser, *J. Mag. Reson.*, 1974, **15**, 44.
[20] D. Ziessow and B. Blumich, *Ber. Bunsenges. Phys. Chem.*, 1974, **78**, 1169.
[21] R. R. Ernst, *J. Mag. Reson.*, 1970, **3**, 10.
[22] R. Kaiser, *J. Mag. Reson.*, 1970, **3**, 28.
[23] J. Dadok and R. F. Sprecher, *J. Mag. Reson.*, 1974, **13**, 243.
[24] R. K. Gupta, E. D. Becker and J. A. Ferretti, *J. Mag. Reson.*, 1974, **13**, 275.
[25] C. F. James and C. L. Wilkins, *Chem. Phys. Lett.*, 1984, **108**, 58.
[26] A. G. Marshall, T. L. Wang and T. L. Ricca, *Chem. Phys. Lett.*, 1984, **108**, 63.
[27] G. R. Haugen, G. M. Hieftje, L. L. Steinmetz and R. E. Russo, *Appl. Spectrosc.*, 1982, **36**, 203.
[28] H. Baba, K. Sakurai and F. Shimizu, *Rev. Sci. Instrum.*, 1983, **54**, 454.
[29] Y. Sugitani, A. Uejimi and K. Kato, *J. Photoacoust.*, 1982, **1**, 217.
[30] G. F. Kirkbright and R. M. Miller, *Anal. Chem.*, 1983, **55**, 502.
[31] G. F. Kirkbright, R. M. Miller, D. E. M. Spillane and Y. Sugitani, *Anal. Chem.*, 1984, **56**, 2043.
[32] T. Arikawa and S. Kaneko, *Mass Spectrom.*, **26**, 313.
[33] T. Arikawa, *Japan J. Appl. Phys.*, 1979, **18**, 211.
[34] T. Arikawa and S. Kaneko, *Japan J. Appl. Phys.*, 1979, **18**, 413.
[35] T. Arikawa, *Electronic and Atomic Collisions*, N. Oda and K. Takayanagy (eds.), p. 799. North-Holland, Amsterdam, 1980.
[36] H. Coufal, U. Moller and S. Schneider, *Appl. Opt.*, 1982, **21**, 116.
[37] R. N. Ibbett, D. Aspinall and J. F. Grainger, *Appl. Opt.*, 1968, **7**, 1089.
[38] J. A. Decker and M. Harwit, *Appl. Opt.*, 1968, **7**, 2205.
[39] M. J. E. Golay, *Opt. J. Soc. Am.*, 1949, **39**, 437.
[40] M. Harwit and N. J. A. Sloane, *Hadamard Transform Optics*, Academic Press, New York, 1979.
[41] R. D. Swift, R. B. Wattson, J. A. Decker, R. Paganetti and M. Harwit, *Appl. Opt.*, 1976, **15**, 1595.
[42] A. B. Giles, *Appl. Opt.*, 1981, **20**, 3068.
[43] S. Miyamoto, H. Tsunemi and K. Tsuno, *Nucl. Instrum. Methods*, 1981, **180**, 557.
[44] N. Sugimoto, *Appl. Opt.*, 1986, **25**, 863.
[45] F. K. Fotiou and M. D. Morris, *Appl. Spectrosc.*, 1986, **40**, 704.

[46] R. Bellman, *Introduction to Matrix Analysis*, McGraw-Hill, New York, 1960.
[47] E. D. Nelson and M. L. Fredman, *J. Opt. Soc. Am.*, 1970, **60**, 1664.
[48] M. Cohn and A. Lempel, *IEEE Trans. Inf. Theory*, 1977, **IT-23**, 135.
[49] A. Lempel, *Appl. Opt.*, 1979, **18**, 4604.
[50] M. Kaljurand and E. Küllik, *Chromatographia*, 1978, **11**, 328.
[51] M. Kaljurand and E. Küllik, *Trends Anal. Chem.*, 1985, **4**, 200.
[52] M. Koel, M. Kaljurand and E. Küllik, *Proc. Estonian SSR Acad. Sci., Chemistry*, 1983, **32**, 125.
[53] N. J. A. Sloane, M. Harwit and M. H. Tai, *Appl. Opt.*, 1978, **17**, 2991.
[54] G. A. Korn, *Random-Process Simulation and Measurements*, Mc-Graw-Hill, New York, 1966.
[55] R. M. Belchamber and G. Horlick, *Talanta*, 1981, **28,** 547.
[56] R. Hoffman, M. M. Gupta and P. N. Nikiforuk, *Proc.IEE*, 1922, **119**, 237.
[57] T. T. Lub and H. C. Smit, *Anal. Chim. Acta*, 1979, **112**, 341.
[58] M. Kaljurand and E. Küllik, *J. Chromatog.*, 1979, **186**, 145.
[59] E. Küllik and M. Kaljurand, *Itogi Nauki Tehniki, Khromatografia*, 1983, **4**, 3.
[60] S. R. Frazer, *Information Extraction in Chromatography Using Correlation Techniques*, Dissertation, Univ. Arizona, Tucson, Arizona, 1985.

3

Two-dimensional measurements involving chromatography

3.1 INTRODUCTION

If two independent parameters are changed during an experiment then we are talking about a two-dimensional method. The best example of this is two-dimensional NMR spectroscopy where two excitation frequences are changed during the experiment (a chemical shift and a spin–spin coupling frequency). In chromatography, one parameter is always the chromatogram running time and in this book it is assumed as one dimension. The other parameter to vary during the measurement is up to the chromatographer. It can be either the carrier gas pressure or the column temperature. The two-dimensional chromatogram is obtained by recording a chromatogram for every carrier pressure or column temperature used. Becoming more popular is column-switching (two-dimensional separation) where the second parameter is also chromatogram running time, but is that for a second column. The greater resolving power allowed by adding a dimension to a separation can actually increase the sensitivity of the technique. This is because the minimum detectable quantity is limited not by the ability to detect a compound, but rather by the presence of other components in the sample under study that affect the detector response, e.g. the detection of small impurities becomes ambiguous when they appear on the tail of a large peak. The ability to spread interfering components out on another column allows the inherent sensitivity of the detector to be used for the component of interest. Besides the chromatographic separation itself there are many interesting possibilities for two-dimensional chromatography in analysing time-varying processes, such as a chemical reaction, an environmental status change, industrial process control, etc.

The best known two-dimensional methods that involve chromatography as one dimension are GC–mass spectrometry and GC–IR spectrometry. However, because these methods generally use only single-injection of a sample we shall not describe them in this book.

Thus let us consider two-dimensional chromatography based on multiple-injection techniques with computer control of the equipment. In its simplest form a number of injections are made in a predetermined sequence. The second parameter is changed for every injection and so a set of chromatograms is recorded. We call this set a 'two-dimensional chromatogram' that can include tens or hundreds of chromatograms. It is clear that recording a two-dimensional chromatogram by manual operation is very tedious, and computer control is needed.

Even so, two-dimensional methods are time-consuming. If we take, for example, two-dimensional NMR, then to record a J-resolved two-dimensional NMR spectrum can take seven hours and another three hours may be needed to compute the spectrum from the experimental data [1]. In chromatography, two-dimensional methods are not so widespread, and it is appropriate to give some general data about recording times. In thermochromatography the experiment usually lasts for 1–2 hours (see Section 5.1.3). If a chromatogram is recorded at n_1 points with a sampling interval Δt between the points, and the second parameter (dimension) is recorded at n_2 points, then the overall two-dimensional experiment time is $n_1 n_2 \Delta t$ and the number of data points is $n_1 n_2$. In a reasonable chromatographic experiment $n_1 = 10^3$ and $\Delta t = 1$ sec. Clearly the requirements for the computer memory and experiment time are quite severe even for n_2 as small as 10^2.

3.2 TIME-RESOLVED CHROMATOGRAPHY AT NON-STATIONARY CONCENTRATION FLOWS

3.2.1 The sampling theorem

Let us first consider the case of two-dimensional chromatography where the second independent variable is the running time (or some function of time) of non-stationary flow changes. Monitoring the composition of time-varying gas and liquid flows requires an instrument with a continuous output. Unfortunately, the latter usually lacks resolving power and the response obtained from a continuous output instrument (e.g. a spectrometer) is a sum of all the individual responses of the species present in the flow. The separation method, chromatography, is a discrete method with respect to time in relation to the process to be analysed, if analysis of the non-stationary flow and the chromatographic separation are performed simultaneously. Separation of the flow components sampled at a particular time takes a certain time in a chromatographic column. The flow composition can be varied during the separation time in such a manner that a second sample can be injected before all the changes of interest in the flow are completed.

Thus quasicontinuous analysis of time-varying processes by chromatography is possible (and also widely used) if the rate of the process and that of the chromatographic analysis are conveniently related. Intuitively, it can be seen that a representative analysis of the composition of the time-varying flow is possible when the variations in the flow are relatively slow compared with the chromatographic analysis. But how 'slow' is slow? The Nyquist sampling theorem [2] (also known as the Kotelnikov theorem [3]) gives an exact answer. Before giving a formulation of the Nyquist sampling theorem we will examine the sampling process in more detail.

Separation of the flow components at a particular time gives the concentrations of the flow components at that moment. If the chromatographic separation takes, say, Δt time units then the next set of concentration values will be obtained after Δt time units. We call this time a sampling interval or the time resolution of the process.

Sec. 3.2] **Time-resolved chromatography at non-stationary concentration flows** 61

When the experiment is complete, a set of digitized concentration values for each flow component is available, i.e. the concentrations as digitized functions of time. The chromatograph acts as a kind of analogue-to-digital converter for the time-varying flow. Now we can consider methods by which a digitized signal (sampled concentrations) can be converted into an analogue signal comparable to the original one.

The method used to regenerate an analogue function from a set of digitized samples can affect the digitization errors. The analogue signal can be approximated by horizontal straight lines drawn from one sample to another. This is the simplest method. A more complicated method could use polynomials to connect sample points. A cubic spline approximation of the sampled data is a very elegant way but requires sophisticated computer programming. Each of these methods will give different errors depending on the properties of the signal to be sampled.

If a signal does not vary according to a simple function (linear or quadratic), then it can be represented by a sum of sine and cosine functions with different frequences. This is performed by a Fourier transform of the signal. If the Fourier transform of the analogue signal $y(t)$ is, say, $F(f)$, where f is a frequency, then the Fourier transform of a digitized variant of the same analogue signal replicates itself at intervals of $1/\Delta t$ (Fig. 3.1). Mathematically this can quite easily be proved, e.g. in the excellent book about the Fourier transform by Bracewell [2].

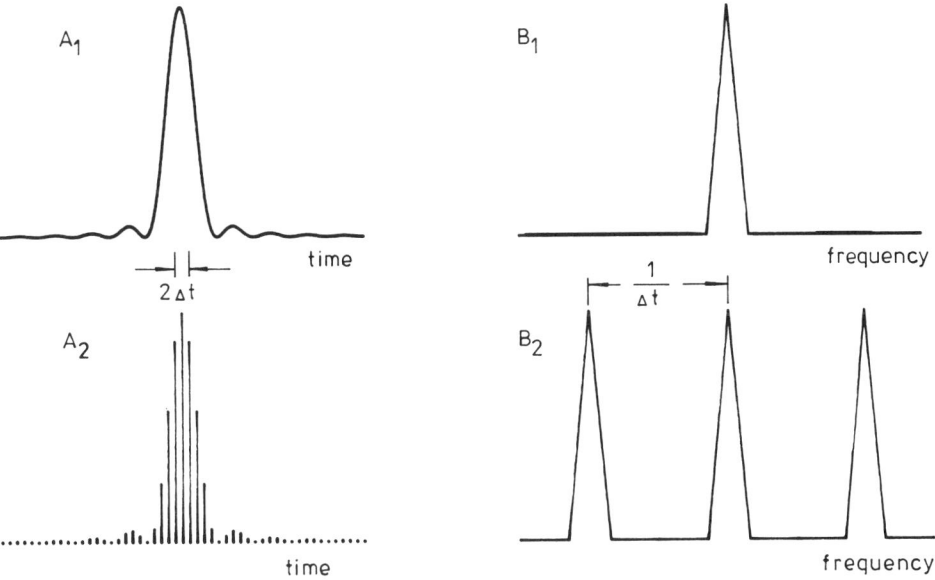

Fig. 3.1 — Illustration of sampling theorem. Upper trace: a continuous function (A1) and its Fourier transform (B1). Lower trace: a digitized function (A2) and its Fourier transform (B2).

Let us consider now how an original analogue signal can be recovered from the sampled data set. The Fourier transform of the sampled data set is multiplied by a box-like function that is abruptly truncated at the point $1/(2\Delta t)$. This generates a function that should be exactly the same as the Fourier transform of the analogue signal, that can now easily be obtained by an inverse transformation of this function (Fig. 3.2).

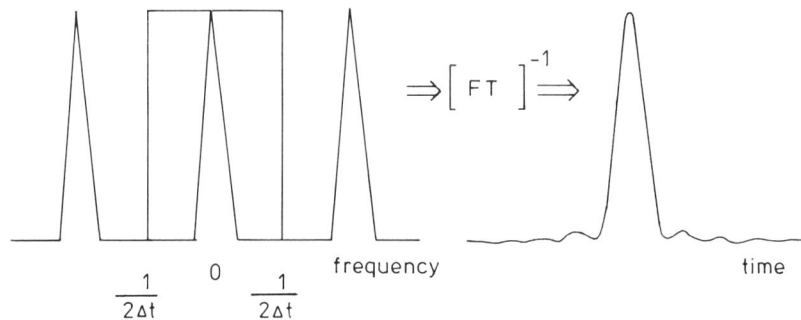

Fig. 3.2 — Reconstruction of a continuous function from its digitized Fourier transform by filtering the latter with a box-like function.

Everything is nice until $F(f)=0$ when $f>1/(2\Delta t)$. This signal is said to be band-limited. If it is not so, then, as follows from Fig. 3.2, an exact recovery of the analogue signal without some error is theoretically impossible because different peaks in Fig. 3.2 overlap. This overlapping is called aliasing. As follows from Fig. 3.1, the reason for aliasing is too long a sampling interval. With decrease in sampling interval the aliasing error decreases.

The sampling theorem can now be formulated as follows: a function for which the Fourier transform is zero for all frequencies $f>1/(2\Delta t)$ is fully specified by values spaced at equal intervals not exceeding Δt.

If this condition of the sampling theorem is fulfilled, then the different peaks in Fig. 3.2 will not overlap and an exact reconstruction of the analogue function is possible. A typical band-limited function is $[\sin(\pi t/7\Delta t)]/(\pi t/7\Delta t)$, the Fourier transform of which is a triangular function.

Of course, in reconstructing the analogue signal from its digitized set of samples it is not necessary to use the Fourier transform. Interpolation by means of polynomials or cubic splines is also possible and is commonly used. The sampling theorem is derived on the assumption that reconstruction is done by convoluting the digital signal with the $[\sin(\pi t/\Delta t)]/(\pi t/(\pi t/\Delta t)$ function. An equivalent procedure is multiplying the analogue signal Fourier transform by a box-like function, as described above. Nevertheless, as the difference between the errors produced by different methods of reconstruction seems not to be great, the sampling theorem gives a good estimation for Δt whatever the method of reconstruction [4].

Although band-limited in principal, the Fourier transforms of real signals often approach zero only at high frequencies, which leads to an unrealistically short sampling interval. It is useful to consider the type and extent of errors resulting when

Sec. 3.2] **Time-resolved chromatography at non-stationary concentration flows** 63

the signal is 'undersampled'. This has been done for some common peak shapes [4]. The results are given in Tables 3.1, 3.2 and 3.3.

Table 3.1 — Definition of peak forms. (Reproduced with permission, from P. C. Kelly and G. Horlick, *Anal. Chem.*, 1973, **45**, 518. Copyright 1973, American Chemical Society).

Peak	Function form (t in sec)
Triangle	$1-|t|$; $|t|\leq 1$ 0; $|t|>1$
Exponential	$\exp[-2\ln2|t|]$
Lorentz	$1/(1+4t^2)$
Gaussian	$\exp(-4\ln2 t^2)$

Note that the half-widths of all the peaks are equal to 1 sec, i.e. the time is measured in units of half-widths.

Table 3.2 — Maximum sampling interval needed for a given accuracy. (Reproduced with permission, from P. C. Kelly and G. Horlick, *Anal. Chem.*, 1973, **45**, 518. Copyright 1973, American Chemical Society).

Maximum error of peak height, %	Sampling interval, sec			
	Triangle	Exponential	Lorentz	Gaussian
10	0.18	0.17	0.54	0.65
1	0.025	0.020	0.28	0.46
0.1	0.0028	0.0022	0.21	0.38
0.01	0.00032	0.00026	0.16	0.33
0.001	—	—	0.12	0.30

Table 3.3 — Minimum number of samples required for a given accuracy. (Reproduced with permission, from P. C. Kelly and G. Horlick, *Anal. Chem.*, 1973, **45**, 518. Copyright 1973, American Chemical Society).

Maximum error of peak height, %	Triangle	Exponential	Lorentz	Gaussian
10	6	20	6 (9)	3(4)
1	40	330	36 (61)	6(7)
0.1	360	4500	150(430)	9(9)
0.01	3200	51000	630	11
0.001	—	—	2600	14

In parentheses in Table 3.3 are a set of results obtained by a different estimation [5]. There is good agreement only for the Gaussian peaks. The numbers in Tables 3.2 and 3.3 are self-explanatory and it follows from these tables and an independent estimation that for a Gaussian peak 5 or 6 points/per peak is the least acceptable number [3]. A simple practical rule is always useful. Taking the Gaussian peak width as 6σ and measuring 12 points on it (precision better than 0.01%) it follows that it is necessary to record 2 points per σ or (because peak half-width $w_{\frac{1}{2}}=2.35\sigma$) 5 points per peak half-width.

The sampling interval is not the only problem to be considered in a digital representation of an analogue signal. The discussion above of the sampling process assumes that samples are taken over an infinite period of time. The effect of taking samples over a finite period of time leads to a truncation error. Information on the part of the signal not sampled is lost. Reconstruction of the analogue signal from truncated digital data leads to the truncation error, which for a peak-like signal tends to be near the truncation points at the peak boundaries, and if sampling is continued until the information loss is insignificant (i.e. it is at noise level), then this error is not a serious problem. Table 3.3 is derived by taking into account both errors (undersampling and truncation).

For the peak-like signals in Table 3.1 the numbers in Tables 3.2 and 3.3 are derived on the assumption that one sample point is taken at the peak maximum. If this is not the case, the errors of undersampling in Table 3.2 increase by a factor of approximately 2 [4].

The other two problems associated with analogue data sampling are the finite length of the digitization time, and variation in the sampling interval. They are not very important in the special case of a 'chromatographic analogue to digital converter'. The digitization is assumed to be fast enough to be considered infinitely fast. It is the time during which the sample is taken from the flow and injected into the chromatograph that is important. If it is not infinitely fast, then the sampling device acts as a kind of low-pass filter in the process under study. This filter smooths the observed signal. The sampling interval variations, i.e. errors in Δt, are usually too small to become a practical problem when modern computer-based timing devices are used to study time-varying processes by chromatography (less than $1\,\mu$sec) because, as proved below, Δt is hardly less than 10 msec.

It follows from this discussion that it is possible to describe and measure continuous gas and liquid flows by chromatography if the requirements of the sampling theorem are fulfilled. Basically, as already pointed out, the duration of the sampling interval is determined by the sample separation time on the column in the case of equi-interval sampling. However, some advanced multisampling techniques (correlation chromatography and stroboscopic sampling) enable time resolution of the observed processes where the duration of the sampling interval is determined by the functioning time of chromatographic injection devices. All these possibilities will be discussed in the following sections.

3.2.2 Equi-interval sampling and high-speed separations

Equi-interval sampling is the simplest form of multiple-sample injection to a chromatograph. The samples are taken from the flow at equal time intervals Δt and separated on the chromatograph during this interval. If the process to follow is too slow, so that $\Delta t > 5$ min, then we are dealing with conventional chromatography. If the chromatogram is known *a priori*, it is possible to reduce Δt by taking into account the carrier gas hold-up time and separation between the peaks. This somewhat tricky technique of injection was used in some early works on continuous chromatography [6,7]. No special devices are needed besides those already available for realizing $\Delta t > 5$ min.

If the condition that 0.5 min $< \Delta t <$ 5 min is required by the sampling theorem for the observed processes, then computer control of the chromatograph and digital data-acquisition become inevitable. The single human operator is not able to realize the required tirelessness, precision and accuracy of timing if $\Delta t < 5$ min. The chromatograph itself needs no particular changes. Modern microcomputer-controlled chromatographs are well suited for working in this sampling interval. A built-in computer can usually control external sampling devices and generate the necessary timing intervals.

Laboratory robots are becoming popular as sample handling devices for chromatographs, although they are too crude to make injections to the chromatograph without special adapters. In some cases autosamplers could be used as sampling devices but they are not as flexible in performance as laboratory robots.

Many interesting processes can be studied within the time resolution of 0.5 min $< \Delta t <$ 5 min. A typical example is the analysis of the evolved gas flow when a sample is continuously heated by a linear thermal programme. This is a complementary method to thermogravimetry (TG). The second independent parameter (dimension) in this two-dimensional method is the temperature. In these experiments the time resolution is usually $\Delta t = 0.5 – 2$ min, and it can be realized by using short capillary columns for separation of the degradation products. The time resolution determines the temperature resolution of the method $\Delta T = b \Delta t$, where ΔT is the temperature interval and b is the heating rate. Thus, by controlling the heating rate it is possible to control the temperature resolution of the method and reach a better compromise between the requirements of the sampling theorem and the separation time of the products. The method we call thermochromatography will be described in Section 5.1.3 more thoroughly. An equipment for computer-aided thermochromatography was described in [8].

For a time resolution of $\Delta t < 0.5$ min a radical reconstruction of the chromatographic apparatus is required. It seems that the lower limit for the sampling interval is $\Delta t = 0.1$ sec; i.e. by using modern chromatographic equipment and column-making technology it is possible to separate some compounds (such as light hydrocarbons) within a tenth of a second.

High-speed gas chromatography was introduced in Golay's first works on capillary chromatography [9] and later, it was developed by Desty [10]. However, the technology of those times was not well suited for realizing high-speed separation. For example, Desty used a rubber hammer to drive the injector piston for fast sample introduction [10]. An oscilloscope was necessary for displaying the chromatogram. After two decades, interest in high-speed chromatography has increased again. Now microprocessors are being used for controlling the sample introduction and data registration. High-speed gas chromatography has been dealt with in several works: Gaspar [11,12], Guiochon [13], Schutjes and Cramers [14,15], Annino [16] and Jonker [17]. Fast separations have also been achieved in liquid chromatography by DiCesare et al. [18].

Fast separation is realized in a short open-tubular column with an inner diameter of 30–50 μm. It has been shown theoretically that the retention time, t_R, of a compound is proportional to the function d^z, where d is the column diameter and z varies between 1 and 2, depending on the column pressure drop [15].

Using short narrow-bore columns is not the only method of realizing fast separations in gas chromatography. Using columns 32 and 250 mm in length, slurry-packed with 10 μm siliceous particles, Jonker obtained on the fastest chromatograms a peak with a standard deviation of 2 msec, at a rate of more than 1.4×10^4 theoretical plates per sec [17]. The advantage of these columns is a larger sample capacity than that of narrow-bore columns.

If complete physical separation of components in the column is not necessary, which is generally the case, information about the qualitative and quantitative composition of the sample can be obtained by mathematical resolution of the digitized detector signals for the component peaks. This idea has been realized in [19]. The sample of interest passes through a short column packed with a molecular sieve and the peaks are detected by a flame-ionization detector. With preliminary information available about sample composition, peak shapes and retention times, it is possible to develop a relatively simple computing method to find peak areas. A flow composition can be analysed by this method within a time resolution of $\Delta t = 15$ sec. However, the need for preliminary information about the flow composition, for computation, restricts the use of this method to monitoring fixed composition flows, i.e. it is possible to follow changes in the quantitative but not the qualitative composition of the flow. The method has been successfully used for controlling industrial processes.

In liquid chromatography, high-speed resolutions are realized within a few minutes (according to the latest book by the staff of the Perkin-Elmer Co. [20]). The equipment components of the high-speed liquid chromatographic system are basically not new. However, the whole system design as such is new, thus permitting performance which was impossible earlier. The new design includes short columns packed with small particles, a small volume detector, and introduction of 1–6 μl samples.

To fully realize the separations achieved in columns capable of high-speed separation, the sample must be introduced in a very narrow band at the beginning of the column and the detector must have a very fast response time. Extra-column effects need special attention in designing a high-speed system. An infinitely narrow band will diffuse in a column according to:

$$N = t_R^2 / \sigma_c^2 \qquad (3.1)$$

where N is the number of theoretical plates, t_R is the retention time of a peak and σ_c^2 is the peak variance, which for the Gaussian peaks means that 99% of the substance is concentrated in a band with a width of $6\sigma_c$.

Band spreading in the injection device and in the detector makes the experimentally registered peak wider than would be expected from Eq. (3.1). In his exhaustive study, Steinberg presents the following expression for the experimentally observed peak variance measured in time units [21]:

Sec. 3.2] Time-resolved chromatography at non-stationary concentration flows

$$\sigma_T^2 = \sigma_c^2 + \sigma_{ec}^2 = \sigma_c^2 + \frac{(\Delta t_s)^2}{12} + \tau^2 + \frac{1}{12}\left(\frac{V_D}{F}\right)^2 \tag{3.2}$$

where σ_T^2 is the experimentally observed variance, σ_{ec}^2 is the sum of extra-column variances, Δt_s is the sampling time during which the sample is introduced onto the column (it is not the sampling interval), τ is the detector time constant (detector response time), V_D is the detector volume, and F is the total gas volumetric flow-rate through the detector. As follows from Eq. (3.2), Δt_s and τ must be of the order of 10–100 msec and V_D in the range 1–10 μl. Then the extra-column contributions to the chromatographic band spreading are comparable to the band spreading in the column. Generally only the flame-ionization detector in GC meets these requirements. The other detectors (e.g. thermal conductivity) need to be specially designed, otherwise they present problems in high-speed separations.

The well-known height equivalent to a theoretical plate (HETP) equation also needs modification in the theoretical treatment of the high-speed system. It was shown that the extra-column contribution to the HETP is proportional to the square of the carrier gas linear velocity, u [12] according to the equation:

$$H = B/u + Cu + Du^2$$

where H is the HEPT value, B and C are coefficients of the classical Golay equation and account for the contribution of the column to the band broadening, while D accounts for the contribution of a finite time of operation of the sampling valve, and a finite volume of the detector and connecting tubing, as follows,

$$D = \sigma_{ec}^2/[(1+k)^2 L]$$

where k is the capacity factor and L is the column length. The extra-column variance σ_{ec}^2 is measured in time units as in Eq. (3.2).

Several sampling devices are in use in high-speed systems. Mechanical valves, activated by compressed helium, can develop switching times of about 7 msec [22]. Fluidic switches can also introduce ultranarrow sample bands during 10 msec. They were introduced into chromatography by Wade [23] and are widely used in high-speed gas chromatographic studies [11–16]. Recently an electrically heated cold-trap inlet system for high-speed gas chromatography was introduced, that can inject a sample in a few msec [24]. The pneumatic-flow switches developed by Deans [25] can introduce samples within 100 msec.

As pointed out above, in high-speed gas chromatography a flame ionization detector is commonly used. It has a low volume, the fluid volume being only a few μl. The detector amplifier is usually modified to decrease the amplifier time-constant. A detector having a time constant of about 1 msec is used in some work. The flame-ionization detector developed for modern capillary chromatography, with a time-constant of about 50 msec, is acceptable in some experiments without any modification. In high-speed liquid chromatography, as mentioned earlier, no special modification of apparatus is necessary, because separation times are generally longer than in the gas chromatographic counterpart.

High-speed chromatography has an important part to play in the measurement of time-varying processes. However, the applications of high-speed chromatography are not restricted to this alone. Owing to its extremely short analysis time high-speed chromatography can greatly increase the throughput for expensive analytical instruments such as mass spectrometers. A mass spectrometer can scan a GC peak very rapidly and if the chromatogram running time approaches an hour, as is common in high-resolution capillary chromatography, the expensive mass spectrometer is not fully used. High-speed GC will provide more samples per unit time and thus utilize the spectrometer more profitably. High-speed separations could also find application in industrial process control, where short separation time ensures a rapid feedback.

Various sampling devices and current equipment available for high-speed chromatography will be described in Chapter 4 in more detail.

3.2.3 Analysis of time-varying gas and liquid flows by correlation chromatography

In correlation chromatography sampling is done at the rate of the detector signal acquisition frequency. The sample component bands severely overlap but the decorrelation operation, as shown in Section 2.2, is able to present a chromatogram in proper form. In the early years of correlation chromatography it was expected that as a sample is taken from the flow at very short intervals, which is also the sampling interval for the chromatogram running time, then quasicontinuous analysis of time-varying flows would be possible [26,27]. However, difficulties arose from the correlation noise. It was proved in Section 2.2.4 that the time-varying input generates some correlation noise on the decorrelated chromatogram. This phenomenon was discovered very early in correlation experiments [28–30]. It is not surprising, because the assumption that the chromatographic system is operated under stationary conditions is not valid when the input function changes with time. Decorrelation generally acts as a kind of low-pass filter and all variations in input will be smoothed and averaged over the length of a chromatogram. These variations are transformed into correlation noise.

In principle, the noise-free decorrelation of the detector output signal is not possible if the amount of sample changes during the correlation experiment. It is mathematically easy to show that because of the non-stationary input, H in Eq. (2.4) can no longer be considered as a vector. The chromatogram is now two-dimensional and its digital presentation is a matrix with elements $H_j(i)$. Here j stands for the chromatogram running time index and i is the non-stationary flow-time index. In other words, as the flow composition changes, a sample taken at the time $k\Delta t$ gives a chromatogram with elements $H_j(k)$, which generally differs from $H_j(l)$ obtained by taking the sample from the flow at the time $l\Delta t$. Here k, l and j are integers. The relationship between column input and detector output can be derived analogously to Eq. (2.3) and is omitted here. The result is:

Sec. 3.2] Time-resolved chromatography at non-stationary concentration flows

$$\begin{bmatrix} Y_1 & \cdot & \cdots & \cdot \\ \cdot & Y_2 & \cdots & \cdot \\ \vdots & \vdots & \cdots & \cdot \\ \cdot & \cdot & \cdots & Y_n \end{bmatrix} = \mathbf{X} \begin{bmatrix} H_1(1) & H_1(2) & \cdots & H_1(n) \\ H_2(n) & H_2(1) & \cdots & H_2(n-1) \\ \vdots & \vdots & \cdots & \vdots \\ H_n(2) & H_n(3) & \cdots & H_n(1) \end{bmatrix} \qquad (3.3)$$

or $\mathbf{Y} = \mathbf{XH}$. In the matrix $\mathbf{H}$ each diagonal is a digital representation of a chromatogram registered at a particular time of flow. Instead of the output vector [as in Eq. (2.4)] Eq. (3.3) gives the output matrix $\mathbf{Y}$. Only the elements of the main diagonal are present and measurable in a correlation experiment. The elements outside the main diagonal of $\mathbf{Y}$ are represented by dots because there are no such expressions as in Eq. (2.3) available for them. Hence, we must still deal with a problem of n equations (the number of the available matrix elements of $\mathbf{Y}$) and n^2 unknowns (the number of $\mathbf{H}$ elements), which has no exact solution. To get an idea of a two-dimensional chromatogram $\mathbf{H}$, some kind of estimation should be made of elements of $\mathbf{H}$.

If the concentration of all the components of the flow change linearly at the same rate, b, then the standard deviation of the correlation noise is given by Eq. (2.22): $s_{\Delta H} = \gamma b \sqrt{n \Delta t}$. Thus, by computing the standard deviation of the correlation noise it is possible to get an estimate of the rate. This conclusion has been supported by various studies [31,32]. The correlation noise form is also important. Although seemingly a random fluctuation of the baseline, it is nevertheless a determinate function of the process rate and chromatogram shape. By examining phase changes in the correlation noise and its value at particular points of the chromatogram it is possible to get information about the location of maxima and minima in the concentration of the flow under study [31].

When each component changes according to a different function, the situation becomes more complicated. By running several correlation experiments in sequence under non-stationary flow conditions it is possible to extend the matrices $\mathbf{Y}$ and $\mathbf{H}$ in Eq. (3.3) in a horizontal direction so that $\mathbf{Y}$ contains several main diagonals. Now, the elements $Y_j(i)$ between the main diagonals can be approximated by some known functions, with the elements of the main diagonal in one row in $\mathbf{Y}$ being used as points to be fitted with this function. This method has been proposed by Hoffman et al. [33] for the estimation of the elements $Y_j(i)$. It is possible to get a reasonable result for flows having linear composition changes.

This conclusion is quite disappointing. An elegant, sophisticated and seemingly powerful correlation method has no advantage over equi-interval sample introduction for time-varying process measurement. For the linearly varying flow there is no problem with satisfying the requirement of the sampling theorem, and equi-interval sampling can be realized by much simpler computational and experimental means. However, the problem has not yet been completely solved. When the process to be studied is so fast and non-linear and the chromatographic separation time severely violates the sampling theorem requirements, then the pseudo-random sampling of the flow and decorrelation of the output signal certainly yields more information about a process than equi-interval sampling. The amount and meaning of this information requires further study.

Correlation chromatography is also superior to equi-interval sampling when the detector signal is noisy. A good example is the study of the evolution of the thermal degradation products of thermostable polymers [34,35]. The gas evolution rate is so low that it is difficult to measure it by conventional chromatographic techniques. The correlation method functions as a noise suppressor and does not decrease the sampling interval for the process. Moreover, the process rate must be quite slow to generate no excessive correlation noise. In thermal analysis, fortunately, process rates can be partly controlled by varying the sample heating rate. Hence, the correlation noise seems to be not a big problem in that kind of study. Equation (2.22) gives an estimation of the expected standard deviation of the correlation noise.

3.2.4 Ensemble correlation and stroboscopic sampling

If the time-varying flow is repeatable, then instead of one correlation experiment it is possible to make several of them (n) for the same process. For each correlation experiment a different PRBS is used. The PRBS for the ith experiment is obtained from the PRBS of the first experiment by cyclically permuting it i times; e.g. if $n = 4$ then the corresponding PRBS are:

$$\{x_1 \ x_2 \ x_3 \ x_4\} = \text{PRBS for the 1st experiment,}$$
$$\{x_4 \ x_1 \ x_2 \ x_3\} = \text{PRBS for the 2nd experiment,}$$
$$\{x_3 \ x_4 \ x_1 \ x_2\} = \text{PRBS for the 3rd experiment,}$$
$$\{x_2 \ x_3 \ x_4 \ x_1\} = \text{PRBS for the 4th experiment,}$$

or in other words, PRBSs from all experiments form the columns of the input matrix **X**. As a result an ensemble of the detector output signals **Y** (n recordings) is obtained. Now it is possible to decorrelate this ensemble not over time but over different experiments, in a similar manner to a conventional ensemble averaging experiment. It can be mathematically formulated analogously with Eq. (2.4) in the following form:

$$\begin{bmatrix} Y_1(1) & Y_2(2) & \cdots & Y_n(n) \\ Y_1(n) & Y_2(1) & \cdots & Y_n(n-1) \\ \vdots & \vdots & \cdots & \vdots \\ Y_1(2) & Y_2(3) & \cdots & Y_n(1) \end{bmatrix} = \mathbf{XH} \qquad (3.4)$$

or **Y** = **XH**. This is the same equation as (3.3) but now the non-diagonal elements of **Y** are determined from experimental measurements. In $Y_j(i)$ the index in parentheses is the number of the experiment and the lower index j is the number of the digitized detector signal element in a particular experiment i.

Now an important point is that all elements in the matrix **H** can be precisely computed without any correlation noise. In this experiment the time resolution of the time-varying process is no longer determined by the chromatographic resolution time but only by the speed of operation of the sampling system and can be as short as 1 msec in a modern sampling system. In fact, this interval determines the shortest time resolution available with contemporary chromatographic systems for time-varying process studies.

Sec. 3.2] **Time-resolved chromatography at non-stationary concentration flows** 71

Ensemble correlation chromatography has also a multiplex advantage over conventional correlation chromatography. This has been confirmed by computer simulation of the method, by the authors. The results are presented in Fig. 3.3. The

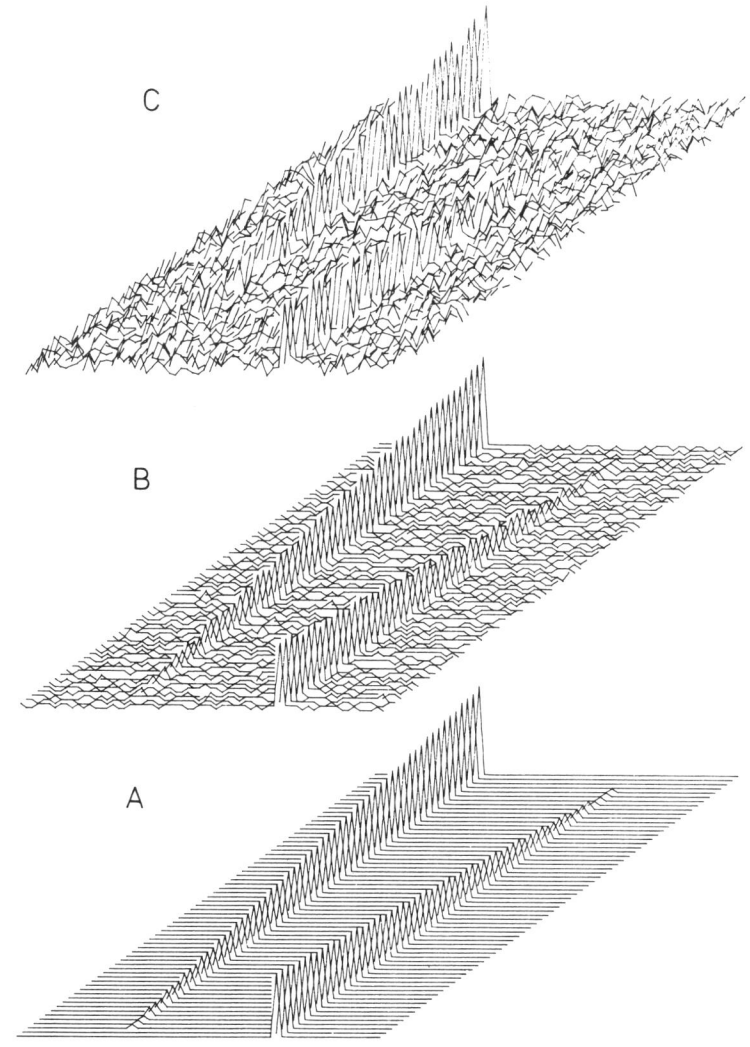

Fig. 3.3—Computer simulation of ensemble correlation chromatography: (A), noise-free two-dimensional chromatogram; (B), two-dimensional correlogram with reduced noise; (C), two-dimensional chromatogram with noise. The detector noise for (B) and (C) has the same power. The PRBS has 63 elements.

simulated process has two components, one increasing linearly in time and the other decreasing. In Fig. 3.3A, a random noise-free output of an ensemble correlation experiment is presented. As can be seen, no correlation noise is present. In Fig.

3.3C, a random noise is added to a two-dimensional chromatogram. The ensemble correlation reduces this noise according to the theoretical rule, as can be seen from Fig. 3.3B. Here $n = 63$.

The authors have no reports available on the application of ensemble correlation, except a short theoretical description [36]. The experimental requirements for ensemble correlation are quite severe. Since the experiment is sensible only in the case of a large number of PRBS repetitions, the sampling system must be able to reproduce the introduction of 10^4–10^6 injections during one ensemble correlation experiment. For mechanical valves this is an unrealistic requirement, but for pneumatic switches controlled by solenoid valves a large number of injections should not be a big problem. The requirements for the computer are also high. It is necessary to compute n^2 elements of the matrix **H**. However, personal computers and especially professional ones equipped with a mathematical processor and megabytes of operative memory are already available and are able to solve computational problems of the ensemble correlation.

The necessary experimental set-up for the ensemble correlation is presented in Fig. 3.4. The computer initiates the process in the reactor (e.g. injecting a reagent)

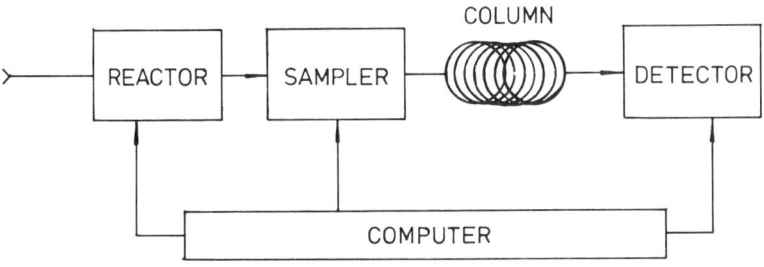

Fig. 3.4 — Apparatus for ensemble correlation chromatography and for stroboscopic sampling.

and controls the PRBS sampling to the column. At the same time the computer records the detector output signal and performs necessary computations. It would be of interest to estimate the time required for the ensemble correlation for some values of n and Δt. If the time resolution is $\Delta t = 10$ msec, and the duration of the process is 10 sec, then $n = 1023$ and the experiment takes three hours. If $\Delta t = 100$ msec, and the process duration is one minute, then $n = 512$ and the experiment lasts for approximately eight hours.

What would be an application of the ensemble correlation method? Chromatographic studies of catalytic processes are becoming increasingly popular [37–39] and of particular interest are on-column reactions [40]. However, the experimental methods used in on-column reaction studies are primitive and out-of-date, and do not offer the necessary time and chromatographic resolution. It seems to us that ensemble correlation chromatography should have good prospects for catalytic studies. Another field in which good time and chromatographic resolution is necessary is the study of ignition and combustion processes. The large number of compounds involved in their reactions complicates the analysis and interpretation of the process. The specification of the behaviour of each component in these reactions would be a good challenge for ensemble correlation chromatography.

Stroboscopic sampling (known also as boxcar chromatography) is another method for analysing repeatable processes with high time resolution. It makes use of the same experimental apparatus as ensemble correlation chromatography but differs in the equipment-controlling program. The process is initiated repeatedly by the computer (see Fig. 3.4) and a sample is injected onto the column only once for each process repetition (in ensemble correlation chromatography the sample is introduced onto the column according to the PRBS during the whole process). By scanning the delay between the initiation and sampling of the process, from zero to the process duration time, it is possible to analyse the whole time-varying flow composition by separating each time the components for a small but every time different part of the process. The boxcar separation of an idealized A→B type reaction chromatogram is presented in Fig. 3.5.

Boxcar chromatography makes fewer demands than ensemble correlation on the sampling valves because the number of injections made during a boxcar experiment is much smaller. The experiment-control, data-acquisition and processing software are not as sophisticated as in ensemble correlation chromatography. However, the latter method has an advantage over boxcar chromatography when low concentration flows are to be dealt with. Stroboscopic sampling has no multiplex advantage.

The applications of boxcar chromatography are the same as those of ensemble correlation chromatography. Snyder reports one variant of the boxcar separation in two-dimensional liquid chromatography with a fixed delay [41,42]. In these works, the time-varying process is the output flow from a liquid chromatograph column through a sampling valve. The process is started by injecting a sample onto the first column. After a preset interval the second sampling valve injects part of the sample separated on the first column onto the second column for further separation. This equipment enables the recording of a total two-dimensional chromatogram. However, the delay between injecting the sample onto the first and second column was not scanned by the authors and thus the separating power of the boxcar method was not fully utilized. The aim of the authors was to separate a pair of peaks of special interest (e.g. for preparative use).

Boxcar chromatography has good prospects for studying on-column reactions, as has been shown theoretically [36]. Injecting a reagent onto a column filled with catalyst results in the so-called catalytic chromatogram which is a chromatogram with wide overlapping peaks of the reagent and products. This overlapping is fundamental, i.e. it cannot be removed by improving the column characteristics. Boxcar chromatography resolves this chromatogram in the second dimension, thus enabling a complete specification to be obtained for each component involved in the reaction. Moreover, reaction rates can be much higher (by a factor of up to 100) than those commonly studied in column reaction chromatography.

3.3 TWO-DIMENSIONAL SEPARATION BY COLUMN-SWITCHING

Switching a solute between columns to obtain better separation between solute components is an old idea dating back to the 1960s. Column-switching was intro-

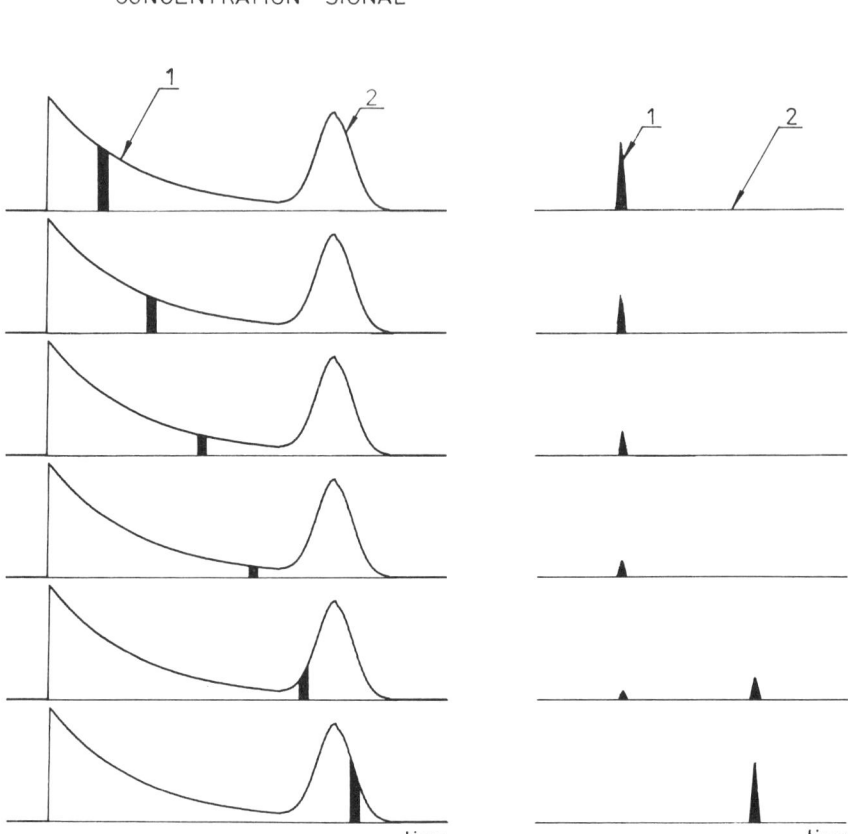

Fig. 3.5 — Stroboscopic sampling of a hypothetical A → B type reaction chromatogram. Peak 1, exponentially decreasing product B; peak 2, reactant A.

duced in the works of Boer [43,44], Huber [45,46] and Deans [47]. However, a rapid increase in the use of column switching can be foreseen only if automatization by means of microcomputers makes this technique easily realizable. Several instrument companies supply apparatus for column-switching [48–51].

This technique is also known under different names: heart cutting (end or front cutting), column-switching, two-dimensional chromatography and multidimensional chromatography. We reject the last term as one leading to confusion and being pretentious. Although this term is widely used in gas chromatography, we know of no experimental multidimensional separations as yet. All these 'multidimensional' separations can be considered as being partly two-dimensional. However, multidimensional separations are possible and have been discussed in the literature from a theoretical point of view [52,53].

Two-dimensional (2D) separations in gas and liquid chromatography make use of two columns with different properties. The first column, where preliminary separation occurs, is usually called a precolumn and the second one, where the main separation takes place, is called an analytical column. The columns are connected by a switching unit and to obtain a really two-dimensional separation they must operate independently (separate carrier gas supply, column temperature control, or eluent composition in LC, and different carrier speed). A typical 2D chromatograph block scheme is presented in Fig. 3.6. According to the nomenclature of Section 3.1, the first column in the 2D chromatograph provides a time-varying input to the second column. Thus, in its most general form 2D chromatography needs two chromatographs.

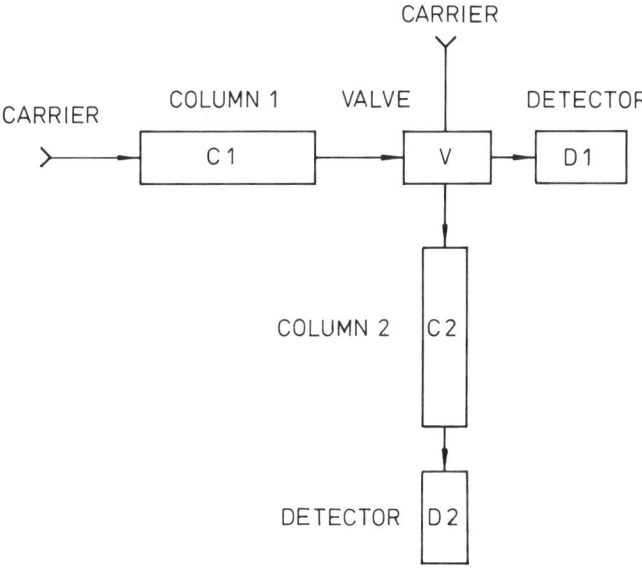

Fig. 3.6 — Scheme of a chromatograph for two-dimensional measurements.

The number of cuts from the first column to the second one is usually small (up to 10) and thus strongly violates the Nyquist sampling theorem. On the other hand, the number of cuts satisfying the sampling theorem will lead to an unrealistically long analysis time and a long computational time. Such an experiment could be very expensive. However, it is not necessary to have the number of cuts suggested by the sampling theorem. Usually we do not want to know everything about the sample. The theorem gives the number of measurements required for the most general signal form. Any *a priori* information about the signal will reduce this number. For example, cuts from the baseline are not necessary. Although trivial, this example demonstrates how prior knowledge about a process can reduce the required number of analyses. The time-varying signal from the first column is in the form of a sequence of peaks and the cuts are usually made from the peaks of interest. In its most sophisticated realization, 2D chromatography utilizes the detector signal from the first column to control sampling to the second column. This is an example of 'intelligent' software control of the instrument, which would help it to be selective about the data that it collects. The operator would specify in advance the particular characteristics of the sample that are to be looked for during the measurement. In Section 3.3.1 we will see how this is realized in liquid chromatography.

When only part of the sample is of interest, the backflush mode of 2D chromatography may be used. This means that after direction of part of the sample of interest to the second column the flow is reversed in the first column. Simultaneously with analysis in the second column the rest of the sample is rinsed out from the first column. In this way the analysis time is considerably shortened when compared with analysis on a single column without backflushing. The reason for this is obvious: it is not necessary to wait for the elution of the compounds with long retention times. Moreover, this method prolongs column life.

3.3.1 Column switching in liquid chromatography

The first suitable column-switching system in which the switching valve could operate at high pressure was proposed by Huber [46]. The system was appraised by allowing a gradual adjustment of the column length and the phase ratio. Below we shall consider some typical case studies involving column switching in LC.

Microcomputer-controlled column-switching systems were used in [54,55]. In these works the switching system does not rely on a predetermined timing sequence but uses a microcomputer to monitor the elution of the chromatographic peaks in real time. The microcomputer makes logical decisions according to a file stored in the computer memory. Automatic column switching systems need very stable chromatographic conditions but the performance of a precolumn is often degraded during an analysis, especially if the sample has had no pretreatment. For this reason the use of computing integrators having an external event relay for controlling different devices has been found inconvenient because they can supply only a fixed timing sequence. Thus, the flexibility of a microcomputer that monitors the detector signal in real time gives a definite advantage over fixed time-sequence devices.

The flexibility is achieved by on-line comparison of the preprogrammed time function stored in the memory with the real time output from the first column detector. In [54] the signal from a UV detector is continuously monitored for a predefined function (e.g. a valley between two peaks) during a given time window Δt_1. Only if both the function and the time window coincide does switching occur. Thus the computer monitors the first column chromatogram run-time and only if the predetermined time window Δt_1 appears does it check the detector signal for a particular prespecified condition during this time window. If the prespecified condition appears on the detector signal (e.g. the beginning of a peak, a valley between peaks or the end of a peak), the computer gives a command to the switching valve for a particular action. In this manner the required bands are transferred from the first column onto another. Prior knowledge of the chromatograms from all the columns is necessary to construct the sequence of time windows and function conditions.

The column-switching technique is very convenient for the group separation of chlorinated dibenzo-*p*-dioxins from most other chlorinated congeners by HPLC.

This group separation is achieved on a microparticulate alumina column. The chromatographic bands of both groups overlap to a high degree on a silica column. However, a silica column resolves chlorinated compounds according to the degree and position of substitution, so after preliminary group separation on the alumina column, the groups can be resolved into their constituents on the silica column. This separation is shown in Fig. 3.7. In this case a critical switching event between peaks 3

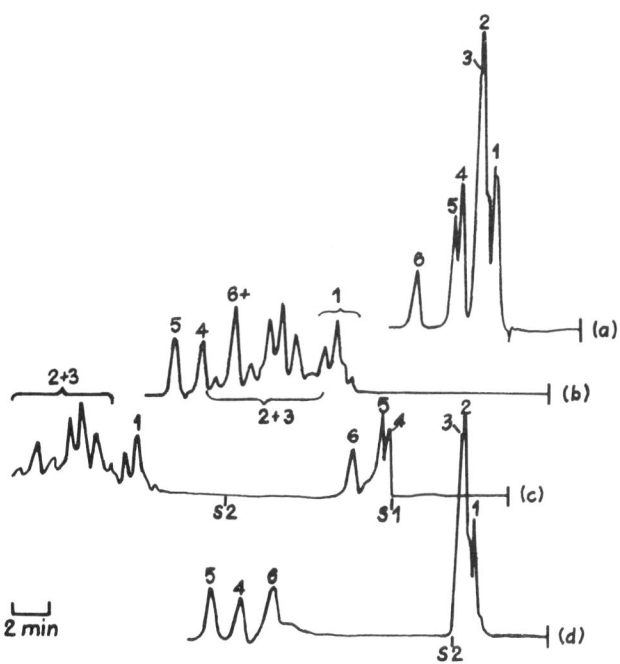

Fig. 3.7 — Separation of dibenzo-p-dioxins from PCBs, p,p'-DDT and p,p'-DDE by column switching. (a) Separation on alumina column only (detector A). (b) Separation on alumina and silica columns in series (detector B). (c) PCBs and p,p'-DDE selectively transferred to silica column for further separation (detector B). (d) Dibenzo-p-dioxins and p,p'-DDT selectively transferred to silica column for further separation (detector B). Peaks: 1,2,3 = PCBs + p,p'-DDE; 4 = dibenzo-p-dioxin; 5 = p,p'-DDT; 6 = octachlorodibenzo-p-dioxin. (After F. W. Willmott, I. Mackenzie and R. J. Dolphin, *J. Chromatog.* 1978, **167**, 31, by permission. Copyright 1978, Elsevier Scientific Publishers).

and 4 was identified by a valley function. Features S1 and S2 determine whether the switching valve between the two columns is directing an eluent from the first column through the second column to the detector, or is bypassing the second column.

Other work demonstrates how organochlorine pesticides in milk were separated from milk fat [55]. Sample clean-up, separation of pesticides and fat determination were done by column-switching and backflushing techniques. Again a sophisticated, microcomputer-generated column-switching sequence was used. In this case, the sequence was fixed in time. In the first period, t_1 (80 sec), fat was retained in the precolumn (50 × 2.1 mm, packed with Partisil 5) and the early-eluting pesticides (from HCB to α-BHC) were led from the precolumn into the analytical one (150 × 3.1 mm, packed with Partisil 10) and stored (trapped) at the end of t_1. These compounds were stored there during the whole of period t_2. Then the flow was switched to bypass the second column. In the period t_2 (680 sec) the late-eluting compounds (e.g. γ-BHC, β-BHC and dieldrin) separated already in the precolumn were led through the bypass, to the detector. The flow was then switched back to the second column and during the period t_3 (225 sec) the early-eluting compounds were separated by the analytical column. At the end of t_3 the valves were switched to the appropriate positions, the primary pump was stopped and the backflush pump was started. The fat was removed from the precolumn with 1% 2-propanol in hexane and determined. Finally the precolumn was regenerated with hexane as solvent. This system describes a fully automated prototype instrument for the analysis of raw milk for pesticide residues. The milk fat was separated from pesticides which in turn were separated. Again precise knowledge of the chromatographic properties of the sample was necessary to enable the desired goal to be achieved. The sample-trapping phenomenon can be exploited by using different solvents in different stages of the 2D system. This is quite analogous to the cryogenic trapping widely used in GC. The preconcentration technique is based on the fact that when relatively non-polar substances in a polar (e.g. aqueous) solution are injected onto a reversed-phase column the substances will be retained in a narrow band at the top of the column. Subsequent elution with a less polar solvent will separate this narrow zone [56]. In fact such a trapping effect was used in determining pesticides in milk, as described above. The trapping effect creates an additional flexibility of separation in 2D systems, giving the analyst the possibility of composing a less time-consuming analysis sequence.

In [57], samples of pharmaceutically active compounds in several biological materials were separated in a 2D system including an RP-2 and an RP-18 column. This system gave excellent preconcentration on the second column. The separation of fluorproquazone in animal feed was achieved by using the following sequence of column switching and mobile-phase composition. During the first few minutes the sample was separated in the first column with a mobile phase of 30% acetonitrile/ 0.01M sodium bicarbonate. The second column was bypassed. After 5.30 min the target band containing fluorproquazone appeared at the end of the first column and the second column was switched on. The target zone was concentrated on the beginning of the second column. After 7.65 min, the first column was switched off and the composition of the mobile phase was changed to 60% acetonitrile/0.01M sodium bicarbonate and the fluorproquazone was separated from the band of interfering components. At the same time, the first column was backflushed. Similar analyses were performed for plasma and urine samples [57]. The total analysis time was very short, reproducibility and recovery were good and the sample preparation was simplified.

In recent years the number of publications dealing with column switching has rapidly increased. Most of them use variants of the basic scheme of coupling the precolumn and the analytical column by a mechanical valve. Valve technology has been improved, allowing the use of quite complicated flow paths with several precolumns or analytical columns [58,59]. In Table 3.4 some applications are presented. These examples have been randomly chosen and are not exhaustive. Most of them utilize similar ideas to those described above and detailed description here would serve little purpose, but they give some idea of possible applications.

Table 3.4 — Application of column switching in HPLC

	Separation or determination problem	Reference
1.	Gradual adjustment of the phase ratio	J. F. K. Huber and F. Eisenbeiss, *J. Chromatog.*, 1978, **149**, 127.
2.	Senna glucosite extract	F. Erni and R. W. Frei, *J. Chromatog.*, 1978, **149**, 561.
3.	Anticonvulsant drug mixture	L. R. Snyder, J. W. Dolan and Sj. Van der Wal, *J. Chromatog.*, 1981, **203**, 3.
4.	Organic trace analysis	H. Hulpke and U. Werthmann, *Chromatographia*, 1979, **12**, 390.
5.	Organic trace analysis	H. Hulpke and U. Werthmann, *Chromatographia*, 1980, **13**, 395.
6.	Herbicides in cereals	J. F. K. Huber, F. Fogy and C. Fioresi, *Chromatographia*, 1980, **13**, 408.
7.	Lower chlorinated aromatic compounds	C. Werkhoven-Goewie, W. M. Boon, A. Praat, R. W. Frei, U. A. Th. Brinkman and C. J. Little, *Chromatographia*, 1982, **16**, 53.
8.	Metoprolol in human plasma	J. B. Lecaillon, C. Souppart and F. Abadie, *Chromatographia*, 1982, **16**, 158.
9.	Lonazolac in human plasma	R. Huber, K. Zeck, M. Worz, Th. Kronbach and W. Voelter, *Chromatographia*, 1982, **16**, 233.
10.	Tetrabutaline in human plasma	L. E. Edholm, B.-M. Kennedy and S. Bergquist, *Chromatographia*, 1982, **16**, 341.
11.	Polynuclear aromatic hydrocarbons	K. Ishibashi, T. Yamasaka and N. Ishibashi, *Anal. Chim. Acta*, 1985, **173**, 169.
12.	Nitro-PAHs in diesel exhaust particulate extracts	W. Lindner, W. Posch, O. S. Wolfbeis and P. Tritthart, *Chromatographia*, 1985, **20**, 213.
13.	Vitamin B in food samples	G. Jaumann and H. Engelhardt, *Chromatographia*, 1985, **20**, 615.
14.	Drugs in plasma	H. Takahagi, K. Inoue and M. Horiguchi, *J. Chromatog.*, 1986, **352**, 369.
15.	Hydroxyproline in meat	A. D. Jones, A. C. Homan, D. J. Fovell and C. H. S. Hitchcock, *J. Chromatog.*, 1986, **353**, 153.
16.	Mefloquine in plasma	P. J. Arnold and O. V. Stetten, *J. Chromatog.*, 1986, **353**, 193.

The main technical problem in constructing column switching equipment is the dispersion of a band in the connecting valves and tubes. The volumes of these should be minimized by any possible means (e.g. by using zero dead-volume fittings and shortening the connecting tubing). When working with commonly used LC column diameters (2–4.6 mm) the available valve production technology is acceptable. In working with micro (0.5–1.3 mm), packed capillary (0.3–0.1 mm) and open tubular columns (0.06–0.01 mm), difficulties arise in the manufacture of mechanical valves with an acceptable volume, thus highly restricting the use of column switching.

3.3.2 Column switching in GC

Column switching in GC is conceptually similar to that in HPLC but the requirements for the switching valves are completely different. The switching valves in

HPLC must work at high pressure and at room temperature. In GC, the switching valves must work at a high temperature but at a pressure only slightly above atmospheric. The mechanical valves used in HPLC are not very well suited for GC conditions because they are prone to leakage and need outgassing.

Deans [25] has solved the problem by using restrictors in the gas flow-path and controlling the gas flows by a solenoid valve located outside the GC oven. The sample does not come into contact with any valves and the system does not involve mechanical moving parts at high temperatures. The only mechanical moving unit, the solenoid valve, can easily be controlled by a timer or a microcomputer.

The principle of a Deans pneumatic flow switch can be explained as follows (Fig. 3.8). The carrier gas enters the system through pressure regulators PC1 and PC2.

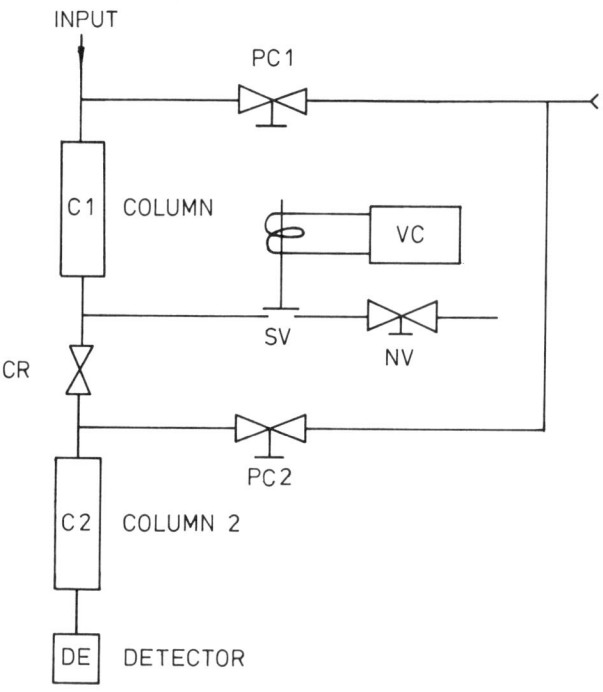

Fig. 3.8 — A Deans-type pneumatic flow switch.

When the solenoid valve (SV) is switched to 'on', the carrier gas and the sample flow through the first column, the needle valve (NV) and then out of the system. Carrier gas is supplied to the second column by regulator PC2, the pressure of which is set slightly above that at the connection between the capillary restrictor (CR) and the second column. This enables the flow through CR to be directed from C2 to C1 and as a result no sample enters the second column. If it is necessary to direct the band of interest into the second column, the solenoid valve is switched to 'off', disconnecting the needle valve from the system, so the carrier gas flows through CR and enables the band to enter the second column. Later, several modifications of the Deans basic flow scheme have been proposed [60–63]. These systems are dealt with in greater detail in Section 4.3.2.

Mechanical valves are also in use in two-dimensional gas chromatography [64,65]. Setting up the flow-rates for the proper operation of the pneumatic switch requires some skill and is extremely difficult when it is necessary to use more than one switch (e.g. in automatic sampling to the precolumn). Although mechanical valves are prone to leakage and have temperature limitations, their distinct advantage over pneumatic flow switches is the independence of the precolumn and analytical column flows, which simplifies adjustment of the carrier gas flow through different parts of the system. There are now new mechanical valves on the market (manufactured by the Valco Instruments Co.) which have an upper temperature limit of 350°C. A careful study of these by the company showed that mechanical valves can be used in two-dimensional gas chromatography if the internal diameters of the valve passages and of the capillary column correspond, and the unswept volumes associated with the fillings are minimized [64].

An important modification in 2D GC was the introduction of a cryogenic trap between the first and second column, by Schomburg *et al.* [66,67]. This trap serves three purposes. First, if a very long duration cut is made, it is necessary to reseparate the chromatographic bands, so part of the sample is collected in the trap and then reinjected into the second column by rapidly heating the trap. Secondly, the trap is necessary for interfacing packed columns to capillary columns in 2D systems. Packed columns require much higher carrier flow-rates than capillary columns and thus the coupling trap is needed for storing solutes. Thirdly, the trap can be used for collection of a particular fraction of the sample from multiple injections into the first column; when enough sample has been collected in the trap for reliable analysis all of it is introduced into the second column by rapidly heating the trap; this is the enrichment mode.

The samples concentrated in the trap are usually introduced into the second column by rapidly heating the trap. According to the latest studies [24] rapid heating permits sample release in only 20 msec, guaranteeing the required narrow initial zone width. This allows the moment of reinjection of the trapped species into the analytical column to be accurately established, as required for the precise measurement of retention indices.

However, problems associated with sample concentration in the trap still remain. They include incomplete recovery of the sample, degradation of sample components either by thermal effects or by reactions between components, irreversible adsorption and micro-fog formation that may result in trap breakthrough. The trap adds extra complexity to the system.

The two-dimensional systems manufactured by instrument companies and described in the literature usually operate automatically under microprocessor control. The SiChromat 2 gas chromatograph [49] can use up to 90 switching commands. A dual oven-and-detector system permits simultaneous detection of chromatograms from both columns. Various analytical data, temperature programs and other experimental parameters can be stored in up to eight files. If an autosampler is connected to a SiChromat 2 it is possible to process a sample fully automatically with eight different analytical and column switching programs [68]. This can be used to optimize column switching conditions or to resolve a complete chromatogram step by step by using shifted cuts, i.e. to utilize the idea of stroboscopic sampling that was described in Section 3.2.4. Sometimes, however, the elaboration of a particular method is limited by the specification of the built-in microprocessor [69].

MUSIC (*MU*ltiple *S*witching *I*ntelligent *C*ontroller) [50] manufactured by the Chrompac Co. has been designed so that it can operate with almost every modern GC. The programmer unit controlling MUSIC enables a step by step analysis of precolumn chromatograms in the analytical column. The time required for step by step analysis is very long because of the long time needed for a chromatogram to be run on the analytical column.

In 2D GC, where time-sequence programming of column switching and trapping is performed by a microcomputer for controlling switching from the precolumn to the analytical column, the analysis of the detector signal is similar to that for HPLC, which in principle is possible with the powerful analytical system described above, and would greatly improve the flexibility of 2D chromatography for a particular analysis.

Recently an interesting study was published in which three different 2D GC systems were compared [70]. Two of these are commercially available: the Scientific Glass Engineering Co. 2D unit (UNIVAP) and the SiChromat. The third system, based on the Valco Co. mechanical valves was home-made. The SiChromat showed an enormous solvent tail which was attributed to adsorption of the solvent on the graphite surface used in the SiChromat pneumatic switching unit connections. The other two systems showed no solvent tailing. Although the precolumn and analytical column were the same for all three systems, the chromatograms obtained with the analytical column for the same switching time and cut length were quite different. The reason for this unusual behaviour was partly due to the complexity of the sample used (tobacco essential oil extract, with an estimated 880 components). However, as no data on the accuracy and precision of the systems compared were published in this work, it is impossible to decide which of the 2D systems was the most reliable. It would be especially interesting to know the relative performances of the mechanical valves and pneumatic switches.

Three examples of the application of 2D GC will now be described. Low concentrations of ethanol in methanol can be determined by overloading the precolumn with the sample and transferring the ethanol-containing band to the second column by heart cutting at the corresponding retention times as specified in the chromatograms in Fig. 3.9 [62]. The overlap due to the methanol tail is removed because the second column is no longer overloaded by the main component and there is also an increase in resolution. The first column is 56 m in length, and the second is 60 m. Both contain poly(propylene glycol) as stationary phase, at column temperatures of 37°C and 47°C, respectively.

The advantages of 2D techniques for the separation of C_1–C_5 hydrocarbons from the products of analytical pyrolysis of coal are demonstrated in Fig. 3.10 [63]. The 2D GC utilizes a low-resolution OV-101 precolumn and a fused-silica porous alumina layer open tubular column. A nice example of the MUSIC performance is given in Fig. 3.11.

Although the importance of 2D GC was pointed out a decade ago [71–73] we still do not see any extensive application of this technique, contrary to the case with liquid

Sec. 3.3] **Two-dimensional separation by column-switching** 83

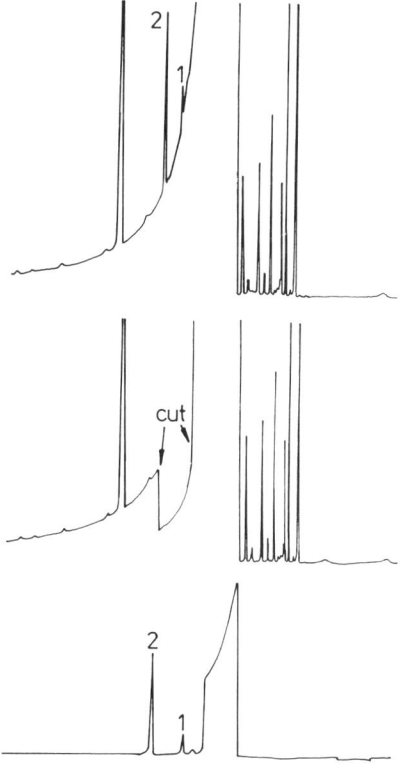

— Separation of traces of ethanol from methanol by heart cutting. Peaks: 1, ethanol; 2, 1-propanol (internal standard). (Adapted from [62]).

chromatography. Several reasons for this have been pointed out by Schomburg et al. [62], as follows.

1. The capillary columns available have high separation efficiency and if temperature programming is used low-volatility compounds can also be dealt with.
2. Most practical analyses are considered to be fast enough without two-dimensional separations.
3. The coupling devices are difficult to design and operate.

It may also be added that 2D systems are quite expensive (since they consist of two chromatographs with a coupling unit). However, if peak overlap in the precolumn is a fundamental problem (e.g. in the case of on-column reactions) there is a good case for using 2D techniques. In recent years the number of publications in the field of 2D GC has somewhat increased. One issue of the *Journal of Chromatographic Science* was completely devoted to column switching in GC [74]. Some recent 2D GC applications are listed in Table 3.5.

As expected, Table 3.5 shows evidence of strong competition between mechanical valves and flow switches in two-dimensional chromatography. The possibility of

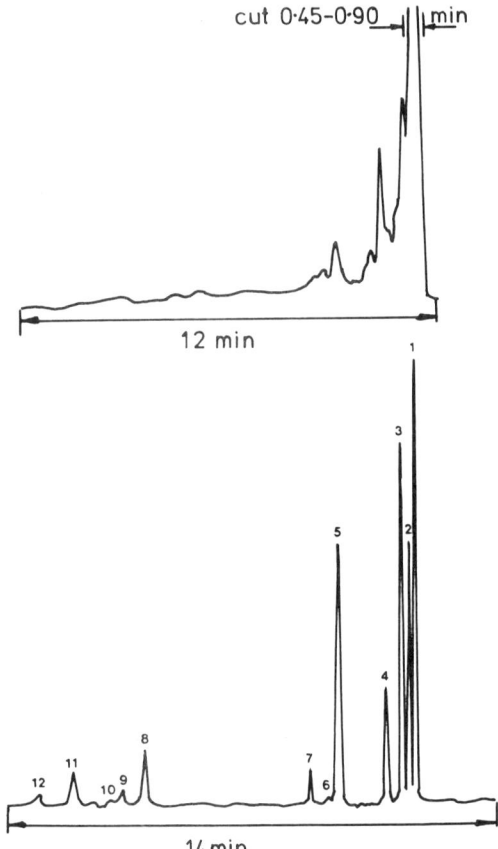

Fig. 3.10 — Two-dimensional separation. Precolumn: 1 m×0.8 mm i.d. stainless steel, 5% OV-101 on Chromosorb W AW DMCS (100–120 mesh). Analytical column: 30 m×0.32 mm i.d. fused-silica Al_2O_3 PLOT. Oven temperature 40°C. Peaks: 1, methane; 2, ethane; 3, ethene; 4, propane; 5, propene; 6, isobutane; 7, butane; 8, 1-butene; 9, *trans*-2-butene; 10, isobutene; 11, isopentane; 12, pentane. (After K. Warman, *J. Anal. Appl. Pyrolysis*, 1984, **7**, 137 by permission. Copyright 1984, Elsevier Scientific Publishers).

solving very complex separation problems by 2D GC is also evident. The problems that have been solved by 2D chromatography include preseparation, sample purification, preconcentration, and heart-cut of key components in complex mixtures, and it has helped in method development and in confirmation of qualitative and quantitative results from packed column GC. These are results which could not have been obtained by using a single column.

3.4 AMOUNT OF INFORMATION OBTAINED WITH TWO-DIMENSIONAL SEPARATION SYSTEMS

3.4.1 Uncertainty and information

Two-dimensional chromatographic systems reveal more information than one-dimensional systems do. A trivial example is thermochromatography (Section 5.1.3)

Sec. 3.4] **Amount of information obtained with 2D separation systems**

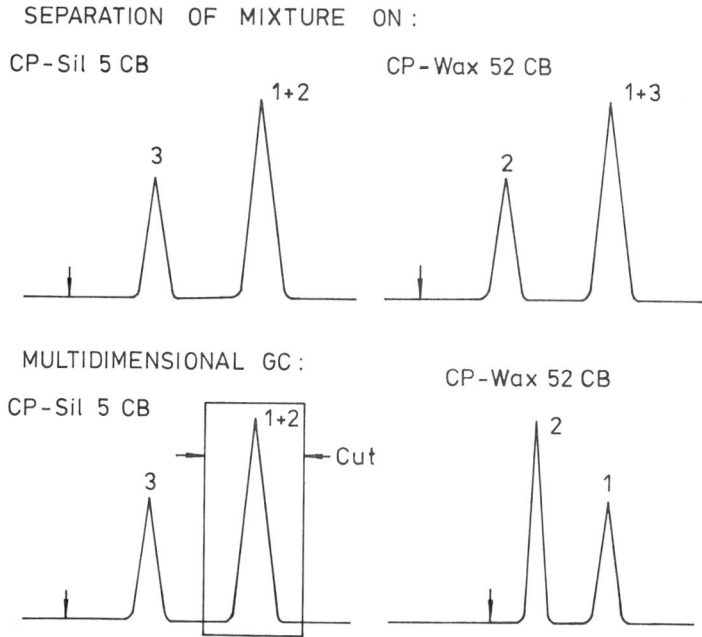

Fig. 3.11 — Separation of hexanal, octane and 2-pentanol by column switching. Peaks: 1, hexanal; 2, octanal; 3, 2-pentanol. (Courtesy Chrompack International B.V., Middelburg, The Netherlands).

which gives the rate of evolution of the sample components with temperature and the retention time of each component (sample component identification). One-dimensional pyrolysis gas chromatography gives only the retention time for each component but the temperature dependence of the evolution rate is lost on the pyrogram. Two-dimensional methods are thus more informative than one-dimensional ones. Information theory can give us an exact measure of the amount of information obtained by performing a particular experiment or analytical measurement.

We can obtain information from an experiment only when the result removes some of the uncertainties existing before the experiment was performed. Thus, there is a relationship between information and the uncertainty removed. One system has more uncertainty than another if it has more states than the other. Let us consider a simple example. A light switch in a bedroom has two different states: 'on' and 'off'. The experiment of testing the switch state may lead to two results with equal probability, $p_{on} = p_{off} = 1/2$. The switch state probability may, of course, be different from 1/2, depending on the time of day. For example, at midnight and midday the 'off' state is more probable and we get less information from our 'experiment' when testing the light switch state at midday than when testing it in the evening. In contrast, an environmental sample may contain 10^3 different species and even testing only for the presence or absence of all these species would require 2^{1000} tests, an almost infinite number. Testing an environmental sample by capillary chromato-

Table 3.5 — Some selected applications of two-dimensional gas chromatography

	Aim and sample	Two-dimensional system	Reference
1.	Determination of nitrosamines	Packed and SCOT columns with two microvolume switching valves	T. A. Gough and K. Sugden, *J. Chromatog.*, 1975, **109**, 265.
2.	Determination of small concentrations of ethanol in methanol	SiChromat 2	G. Schomburg, F. Weeke, F. Müller and M. Oreans, *Chromatographia*, 1982, **16**, 87.
3.	Naphtha and wine volatiles analysis	Packed and capillary column with Deans-type home-made flow switch	R. J. Phillips, K. A. Knauss and R. R. Freeman, *HRC&CC*, 1982, **5**, 546.
4.	Hydrocarbon group analysis in gasoline and diesel fuels	LC and GC coupling	J. A. Apffel and H. McNair, *J. Chromatog.*, 1983, **279**, 139.
5.	Determination of about 100 halogenated and organophosphorus pesticides	SiChromat 2 GC with two capillary columns	H.-J. Stan and D. Mrowetz, *J. Chromatog.*, 1983, **279**, 173.
6.	Determination of chlorinated hydrocarbons: PCP, PCDD and PCDF	SiChromat 2 GC with two capillary columns	G. Schomburg, H. Husmann and E. Hübinger, *HRC&CC*, 1985, **8**, 395.
7.	Analysis for phenanthrenes in crude oil	SiChromat 2 GC with two capillary columns	G. Schomburg, F. Weeke and R. G. Schaefer, *HRC&CC*, 1985, **8**, 388.
8.	Determination of polychlorinated biphenols in coal tar	Packed capillary LC and capillary GC with 10-port Valco valve	H. J. Cortes, C. D. Pfeiffer and B. E. Richter, *HRC&CC*, 1985, **8**, 469.
9.	Synthetic fragrance mixture	Two capillary columns with 6-port Valco valve and FI-IR detector	S. L. Smith, *HRC&CC*, 1985, **8**, 385.
10.	Target compound determination of low-level impurities in styrene monomer and crude oil	Packed and capillary columns, vacuum-induced flow switch and MS detection	W. V. Ligon, Jr. and R. J. May, *J. Chromatog. Sci.*, 1986, **24**, 2.
11.	Pyrolysis GC product profiling	Two capillary columns and SGE flow switch (UNIVAP)	D. W. Wright, K. O. Mahler, L. B. Ballard, and E. Dawes, *J. Chromatog. Sci.*, 1986, **24**, 13.
12.	Organoleptic evaluation of an orange extract	Two capillary columns and SiChromat 2 chromatograph with 'live' switch	P. A. Rodrigues and C. L. Eddy, *J. Chromatog. Sci.*, 1986, **24**, 18.
13.	Analysis for PCBs in methylene chloride	Two capillary columns and Valco 6-port valve	S. Sonchik, *J. Chromatog. Sci.*, 1986, 24, 22.
14.	Simplification of identification of tobacco smoke components	Two capillary columns and Valco 4-port valve with MS detection	J. F. Elder, Jr., B. M. Gordon and M. S. Uhrig, *J. Chromatog. Sci.*, 1986, **24**, 26.
15.	Natural and refinery gas analysis	Macrobore WCOT and PLOT columns with 6-port and 10-port Valco valves	Z. Naizhong and E. L. Green, *HRC&CC*, 1986, **9**, 400.
16.	Naphtha analysis	Packed and capillary columns with Chrompack MUSIC flow switch	*Chrompack News*, 1985, **12**, 1.
17.	Hydrocarbons and inert gases	6-Port mechanical valve and two pneumatic valves (SGE) under control of NEC PC 9801	H. Tani and M. Furuno, *HRC&CC*, 1986, **9**, 712.

graphy gives a tremendous amount of information in a single experiment, far more than is obtained by a single test of the state of a light switch.

Let us consider an experiment with a result that is unknown beforehand, but for which we know all the possible results and their probabilities $p_1, p_2 \ldots p_n$. Taking

Sec. 3.4] **Amount of information obtained with 2D separation systems** 87

into account the number of states of the system and the probability of each state, the amount of uncertainty in a particular system can be calculated according to the Shannon formula:

$$H = -\sum_{i=1}^{n} p_i \log_2 p_i \qquad (3.5)$$

Here H is the entropy, a measure of the system uncertainty, p_i is the probability of a particular state of the system and n is the number of different states that are possible. The information is equal to the uncertainty removed. It is obvious that because of random error in a real experiment we cannot remove all the uncertainty in performing an analytical measurement. The amount of information I given by an experiment can be written as [75]:

$$I = H - H' \qquad (3.6)$$

where H is the degree of uncertainty before the experiment and H' the degree of uncertainty after it. H' is a measure of the noise. In the following discussion, however, we assume that H' is zero.

A unit of information is called a 'bit'. The testing of a two-positional switch with equal probabilities $p = 1/2$ for the two switch states gives $H = -0.5 \log_2 0.5 - 0.5 \log_2 0.5 = 1$ bit of information. At midnight the light switch probabilities could be say $p_{off} = 7/8$ and $p_{on} = 1/8$ and the amount of information is then $I = -(7/8) \log_2 (7/8) - (1/8) \log_2 (1/8) = 0.54$, which is indeed less than that obtained by testing the light switch with equal state probabilities.

This trivial example demonstrates well how commonsense ideas about information may be transformed into quantitative values. In the next section we will see that a chromatogram can be regarded as a connection between two information states. The amount of information obtained from a chromatogram can easily be calculated.

3.4.2 Chromatographic peak capacity and the amount of information obtained by column chromatography

The informational criterion for characterizing the quality of performance of a chromatographic column will now be introduced. Let us consider two columns, one bad, the other good, i.e. the first column has a small number of theoretical plates and the other a large number. Only a few peaks would be resolved with a bad column, whatever the composition of the sample. The good column could resolve very complicated mixtures, giving a large number of peaks on the chromatogram. The peak capacity, Q, or the maximum number of peaks that can be resolved completely between the non-retained component peak and the most highly-retained solute peak is a useful criterion for calculating the amount of information it is possible to obtain from a chromatographic column (Fig. 3.12). Let $Q = 2$ for a bad column and $Q = 100$ for a good column. The column with $Q = 2$ can show only two resolved peaks, so only four different types of chromatogram can be obtained with this column, assuming

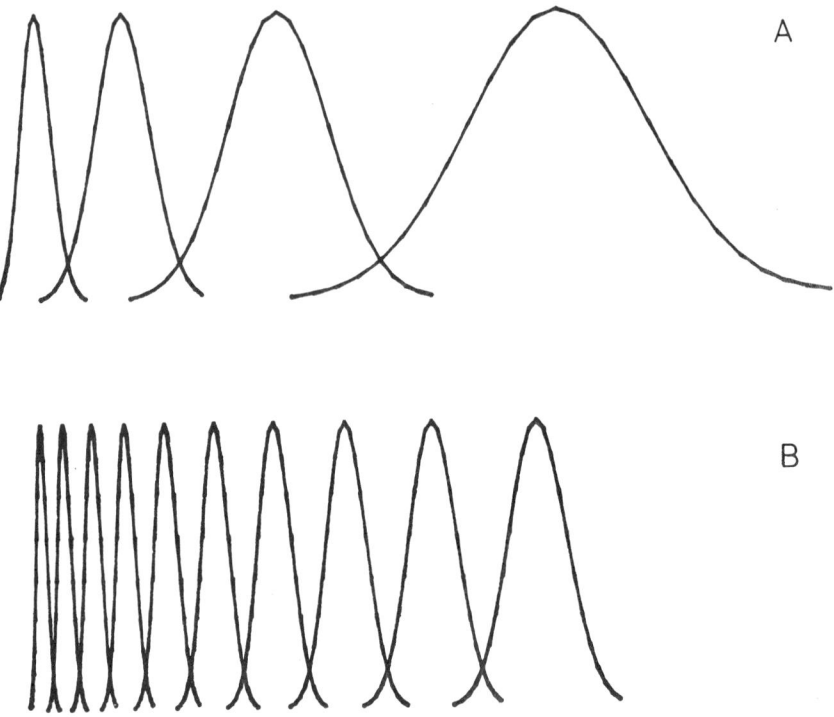

Fig. 3.12 — Location of maximum number of peaks on chromatogram: (A) 'bad' column; (B) 'good' column.

that there is equal probability for the presence and absence of the peaks. Let '1' denote the presence and '0' the absence of a peak on the chromatogram. Then these four chromatograms are (00), (01), (10) and (11). Hence only 2 bits of information can be obtained with the 'bad' column. With the 'good' column 100 bits of information could be obtained because 2^{100} different types of chromatogram are possible with it. Here we do not consider the intensities of the peaks. It is readily seen, from an information theory point of view, that a chromatogram can be considered as a connection between two two-state units.

The term peak capacity was first introduced by Giddings [76,77] and also used by Grushka [78] in theoretical studies of chromatographic column performance. An equation for computation of the peak capacity can be derived as follows. First we assume that the number of theoretical plates, N, is the same for all solutes. This assumption is only approximately valid, but the number of plates will not usually vary from one solute to another by more than 20% [78]. The relationship between the ith peak retention time t_i and the amount of band spreading expressed by the variance σ_i^2 of a Gaussian peak is given by Eq. (3.1): $\sigma_i = t_i/\sqrt{N}$. Let us assume that all the peaks are separated. The condition for separation of adjacent peaks is $t_{i+1} - t_i = 2(\sigma_{i+1} + \sigma_i)$. Using Eq. (3.1) to eliminate the variances σ_{i+1}^2 and σ_i^2 we get

Sec. 3.4] Amount of information obtained with 2D separation systems

$$t_{i+1} = \left(\frac{1+2/\sqrt{N}}{1-2/\sqrt{N}}\right) t_i$$

and in general, for the Qth peak

$$t_Q = \left(\frac{1+2/\sqrt{N}}{1-2/\sqrt{N}}\right)^{Q-1} t_0 \qquad (3.7)$$

where t_0 is the retention time of the first peak in the chromatogram (the carrier gas hold-up time). Taking natural logarithms, we can express Q as

$$Q = 1 + \frac{\ln(t_Q/t_0)}{\ln(1+2/\sqrt{N}) - \ln(1-2/\sqrt{N})}$$

and because $2/\sqrt{N} \ll 1$, by using the approximation $\ln(1+x) = x$ for $x \ll 1$, we finally obtain

$$Q = 1 + \frac{\sqrt{N}}{4} \ln(t_Q/t_0) = 1 + \frac{\sqrt{N}}{4} \ln(1+k_Q) \qquad (3.8)$$

where k_Q is the capacity factor of the Qth solute. Equation (3.8) was first reported by Grushka [78]. The peak capacity can be improved by keeping the width of the peak constant by temperature programming in gas chromatography or by gradient elution in liquid chromatography. The retention time t_f of the last component is:

$$t_f = t_0 + (Q-1)4\sigma$$

from which $Q_{TP} = 1 + \frac{\sqrt{N'}}{4}(t_f/t_0 - 1) \qquad (3.9)$

where $\sqrt{N'}$ is the number of plates for the first (inert gas) peak and Q_{TP} is the peak capacity when temperature programming is used. The improvement in the peak capacity by use of temperature programming in GC (or gradient elution in LC) is due to the fact that Q_{TP} depends linearly on t_f/t_0 and not logarithmically as in Eq. (3.8).

The amount of information that it is possible to obtain by using a particular column having a peak capacity Q is simply

$$I = Q \qquad (3.10)$$

This result is valid under the following assumptions:

(1) a chromatogram is divided into Q sections with increasing width according to Eq. (3.1);
(2) each solute peak can be located in one of these sections;
(3) the probability that a particular solute peak can be located in a section of the chromatogram is equal for all sections;
(4) all peaks have the same intensity.

In this manner a chromatogram is regarded as a set of two-state units, one state being the absence of the peak and the other its presence in that unit of the chromatogram. If the peak capacity is Q, then there are 2^Q different states of the chromatogram, each of equal probability 2^{-Q}. Hence Eq. (3.10) follows directly from Eqs. (3.5) and (3.6).

From the point of view of practical chromatography, assumptions (1)–(4) may seem too broad. However, a peak appearing on the border of two adjacent units can be assigned to either of them. According to some studies, the distribution of the peaks on chromatograms seems to be not uniform but Gaussian [79], but few data are available about this. The peak intensities are important in deriving a general equation for the amount of information obtainable from a chromatographic measurement. The detector characteristics must also be taken into account. For column characterization, where only the separation characteristics are of interest, peak intensities are not so important.

3.4.3 Amount of information obtained with a two-dimensional system

Having two successive columns with different polarities and peak capacities, a 2D system has $2^{Q_1 Q_2}$ different states. The amount of information obtained with this system is:

$$I_{2D} = \sum_{i=1}^{2^{Q_1 Q_2}} 2^{-Q_1 Q_2} \log_2 2^{Q_1 Q_2} = I_1 I_2 \tag{3.11}$$

i.e. multiplication of the amounts of information I_1 and I_2 obtained with the two columns, each regarded as an independent one-dimensional system. From Eq. (3.8) we get

$$I_{2D} = \left(1 + \frac{\sqrt{N_1} \ln (1 + k_1)}{4}\right) \left(1 + \frac{\sqrt{N_2} \ln (1 + k_2)}{4}\right) \tag{3.12}$$

where subscripts 1 and 2 stand for the first and the second columns respectively.

The difference in polarity between the two columns is important. As follows from Eq. (3.8) the value of the polarity (expressed by the ratio t_Q/t_0) contributes only a little to the numerical value of the amount of information. This is because of the logarithmic dependence of the amount of information on t_Q/t_0. However, the

Sec. 3.4] Amount of information obtained with 2D separation systems

difference in polarity allows us to assume the columns to be independent (at least on the assumptions made in Section 3.4.2). Hence a solute peak can have any position in the second column chromatogram independently of its position in the first column chromatogram. The peak capacity value for the combination of both chromatograms is thus the product of the peak capacity values of the two chromatograms [52]. This is expressed in Eq. (3.11). Without the difference in polarity, the combination of the two columns can only be regarded as a single column having approximately $\sqrt{N} = (\sqrt{N_1} + \sqrt{N_2})/\sqrt{2}$ plates but a much lower number of possible states, and thus much less information can be extracted from a chromatogram obtained with this column.

Let us clarify this statement by a simple example. If $N_1 = N_2 = N$ and $N \gg 1$ and $k_1 = k_2 = k$ then $I_{2D} \approx N[\ln(1+k)]^2/16$. So, if the amount of information obtained in a one-dimensional method is proportional to the square root of the plate number the amount obtained in a two-dimensional method changes linearly with the plate number [52,79]. A certain separation problem requires, say, $N = 10^6$ plates on a good capillary column. In the 2D system this separation requires only approximately 10^3 plates in each column, i.e. conventional packed columns in both stages. This estimation may seem too rough, but it demonstrates the power of two-dimensional methods in terms of information content.

Another example is based on published chromatograms. In [80], a capillary column chromatogram of wine volatiles is given together with analysis of the same sample in a 2D system which consisted of a packed column and a capillary column. The chromatogram in the 1D system had a peak capacity $Q = 100$. The packed column chromatogram in the 2D system had a peak capacity of $Q_1 = 22$ and the capillary column chromatogram had $Q_2 = 90$. Thus the 2D system had a total peak capacity of about 2000. These figures are quite approximate owing to difficulties in measuring the peak widths from the published chromatograms. As temperature programming was used in these experiments, Eq. (3.9) was applied for the peak capacity estimations. Only certain parts of the chromatograms were considered in the computation.

As pointed out, the definition of peak capacity is somewhat approximate and arbitrary, and this also applies to the amount of information that it is possible to obtain with a column. By use of different assumptions other measures of the amount of information are possible. The results will of course be different from those obtained by using Eq. (3.12). Sevčik [79] used the following formula for the information content of a chromatographic experiment on a particular column

$$I(s) = \log_2 \left[\frac{N_{\text{eff}}}{4} \frac{(A-1)}{A} \right] \qquad (3.13)$$

where $I(s)$ is the information content, N_{eff} is the effective number of theoretical plates and A is a measure of the column polarity $A = (t'_{n+1} - t'_n)/(t'_n - t'_{n-1})$ where t_{n-1}, t_n, t_{n+1} are the adjusted retention times for three neighbouring n-alkanes. According to Sevčik the information content is based on the resolution between the neighbouring n-alkanes. Despite the difference between Eqs. (3.8) and (3.13) the conclusion is the same: the amount of information obtained in a chromatographic analysis depends on the efficiency and polarity of the column used. Sevčik also considered different modes of column switching. For example, he showed that the combination of a short precolumn with $N_1 = 10^3$ plates and a short capillary column with $N_2 = 2 \times 10^3$ plates is comparable to a single capillary column with $N = 10^6$ plates.

With similar assumptions to those presented in Section 3.4.2, Huber et al. [81,82] derived another formula for the peak capacity:

$$Q = 1 + \ln\left(\frac{t_Q}{t_0}\right)/\ln(1 + R/\sqrt{N}) \approx \frac{\sqrt{N}}{R} \ln\left(\frac{t_Q}{t_0}\right)$$

where R is the desired resolution and N is the number of theoretical plates. It is easily seen that Huber's result does not differ much from Gidding's formula, Eq. (3.8). Huber demonstrated from experimental data that in the sense of the amount of information obtainable, two-dimensional gas chromatographic data can compete with binary-coded mass spectra for identification of the doping drugs.

The ideas of this section can be summarized in a few words: for a particular retention time interval many more different chromatograms are possible in the 2D system than in a 1D system, and the amount of knowledge obtainable with the 2D technique is much larger. Equation (3.12), giving the amount of information, is a measure of this knowledge.

It follows from this discussion that if the required amount of information for a certain separation has been predetermined then the performance requirements for the components of a 2D system are also predetermined and are significantly lower than those for a 1D system. However, it is important to note that these requirements refer only to the column performance. The two-dimensional system also includes sampling devices and data processing units, and is much more complicated than the one-dimensional system. If a 1D system cannot supply the necessary amount of information then the 2D system has a definite advantage, but when the required amount of information can be obtained by both systems then this advantage is lost, and the choice of system depends on the cost and degree of sophistication of the columns as well as on the sampling system.

3.5 SELECTIVITY TUNING

Recently the continuous tuning of the polarity of two columns coupled in series has attracted attention [83–87], although it is an old idea (see literature in [83]). The technique permits predictable positioning of peaks as well as determination of the retention indices and their temperature-dependence. The method is also known as 'multi-chromatography' [84]. Selectivity fine tuning is realized in a fixed-length ('tandem') system of two columns coupled in series, and the selectivity is tuned by adjusting the residence time of the sample components in the individual columns, by changing the carrier gas velocities or the column temperatures independently. The columns of the tandem system can be coupled in series either directly or through a

Selectivity tuning

switching device, and can be mounted either in a single oven or in an independently controllable pair of ovens.

When the selectivity for a solute is fine-tuned by changing only the carrier gas velocities, multi-chromatography requires an independent pressure source between the two columns (Fig. 3.13). The corresponding theory was developed by Hindshaw

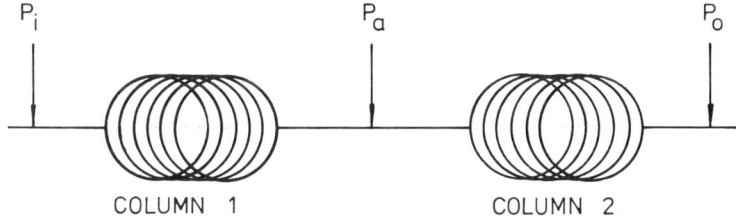

Fig. 3.13 — Basic scheme for the tandem two-column system used for selectivity tuning. p_i, p_a and p_o are the input, midpoint (auxiliary) and output pressures.

and Ettre [83]. Let us denote the variables for the total system and the first and second column as follows: gas hold-up time: t_{ms}, t_{m_1}, t_{m_2}; retention time: t_{Rs}, t_{R1}, t_{R2}; relative retention time: t'_{Rs}, t'_{R1}, t'_{R2}; capacity factor: k_s, k_1, k_2. Because $t_{ms} = t_{m1} + t_{m2}$, $t_{Rs} = t_{R1} + t_{R2}$ and $k = t'_R/t_m$, we obtain:

$$k_s = (t_{m1}/t_{ms})k_1 + (t_{m2}/t_{ms})k_2$$

and because $t_{m2}/t_{ms} = 1 - t_{m1}/t_{ms}$,

$$k_s = (t_{m1}/t_{ms})(k_1 - k_2) + k_2$$

Thus, by varying the ratio t_{m1}/t_{ms} from 0 to 1, the capacity factor of a solute can be changed between k_1 and k_2. As pointed out above, varying the ratio t_{m1}/t_{ms} can be achieved by varying the midpoint pressure p_a. For equal-length columns, and ignoring the carrier gas compressibility, the total gas hold-up time is the sum of the hold-up times of both columns, and is proportional to $t_{ms} \sim 1/(p_i - p_a) + 1/(p_a - p_o)$, from which we obtain

$$t_{m1}/t_{ms} = \frac{p_a - p_o}{p_i - p_o}$$

where p_i, p_a, p_o are the inlet, midpoint and outlet pressures of the tandem system. Thus the total capacity factor k_s can be varied between k_1 and k_2 by changing the midpoint pressure between p_i and p_o.

Selectivity tuning leads to a powerful computerized two-dimensional and multi-input chromatographic technique. It was demonstrated by Kaiser *et al.* [84,85]. The computer controlled several consecutive injections with the same sample at different midpoint pressures. The pressure-adjusting system was a digitally controlled piezo-electric fluidic system that controlled the flow with μl/min precision. In these experiments, a retention index change of several hundred units was found for polar test compounds, such as phenols, with an SE-30/Carbowax tandem system.

After a series of runs a two-dimensional output was obtained, the variables being the chromatogram running time and the midpoint pressure between the two tandem columns. This 2D output could be used for analytical characterization of the sample. From the 2D output optimum polarity conditions could also be derived for further routine analysis. The principle is very popular in LC but as yet has received little attention in GC. In LC a series of chromatograms is recorded with different eluent compositions and the quality of separation is estimated from the values of various functions computed from the chromatogram (see e.g. [88]). For the next sample, the chromatographic conditions are derived from the quality of the separation function, with the help of a suitable algorithm (e.g. simplex optimization [89]). The only example we know of optimizing separation in GC was a somewhat trivial example of optimizing the carrier gas velocity and temperature in a single column [90]. The tandem systems are a real challenge to the simplex optimization method because at least three different parameters can be varied during the optimization process: the pressure between the columns, and the temperatures of the first and second columns.

Selectivity tuning should not be confused with a 2D separation, in which preselected solutes are selectively removed from one column to the other. In selectivity tuning, all the solute passes through both columns, which act as a single column with a selectivity which is a composite of the selectivities of tandem component columns. Evidently there is a limited possibility of using 2D separation for selectivity tuning and *vice versa*.

3.6 PRESENTATION OF TWO-DIMENSIONAL CHROMATOGRAMS

Presentation of 2D data requires special attention. For a 1D system, data presentation is simply a chromatogram on the chart recorder or, in more modern equipment, on a video display. A two-dimensional chromatogram can be considered as a surface in a three-dimensional space. Imaging a three-dimensional object in a plane is the purpose of computer graphics science [91] and is beyond the scope of this book. Here we will consider only the three most popular forms of plotting, that can easily be realized by microcomputer, e.g. on an Apple, and are widely used in analytical data presentation, *viz*. isometric projection, slices of a surface, and contour plots.

3.6.1 Isometric projection

Isometric projection (also called a stack plot, (see Fig. 5.23, p. 178) is a presentation of a 2D chromatogram giving the impression of a three-dimensional image. This projection can be constructed by means of a microcomputer with the following algorithm.

Let us consider a digital 2D chromatogram $H_j(i)$ in which the indices i and j are determined as usual: j is the chromatogram running time and i is the running value of the second variable. Let us assume that the variable j is plotted on the x axis and i on the y-axis of the plotter xy-plane and the plotting increments are Δx and Δy correspondingly. Let us assume that $\mathbf{H}_j(i)$ is normalized so that all its elements are positive and their maximum value does not exceed that acceptable by the plotter. The plotting algorithm contains the following steps:

(1) The first chromatogram $\mathbf{H}_j(1)$ is plotted on the bottom of the xy-plane, starting from the lowest left corner of the plotter ($x = 0$, $y = 0$).
(2) The second chromatogram is plotted above the first. The distance between the base lines of the first and second chromatogram is Δy. The plotting of the second chromatogram $\mathbf{H}_j(2)$ begins from the point $x = \Delta x = \Delta y \tan \phi$, where $\tan \phi$ is the angle between the y-axes and the surface projection, $y = \Delta y$.
(3) The nth chromatogram is plotted above the $(n-1)$th chromatogram and the plot starts from the plotter co-ordinates $x = (n-1)\Delta x = (n-1)\Delta y \tan \phi$; $y = (n-1)\Delta y$.

In this algorithm the use of several plotting angles is restricted, and the ϕ value depends on the number of points in both dimensions as well as on the size of the particular plotter in use. To obtain more plotting freedom a more complicated algorithm could be used, but when the plotting time is of prime importance the performance of the algorithm above is quite acceptable.

Some parts of the 2D chromatogram surface usually remain hidden if looked at from a certain direction. These hidden lines can be removed from the picture in the following way.

(1) A comparison vector with the elements V_j is formulated, the initial values for the elements of this vector being zero for all j.
(2) The plotter moves from a particular chromatogram point $(j-1)\Delta x$, $(i-1)\Delta y$ to the next one $j\Delta x$, $i\Delta y$ with the pen lowered if the condition $H_j(i) > V_j$ is valid. In this case, V_j obtains a new value $V_j = H_j(i)$. The line is drawn in the plotter plane.
(3) The plotter moves from a particular chromatogram point to the next point with the pen lifted if the condition $Hj(i) \leq V_j$ is valid. In this case V_j remains unchanged in the ith point.

This algorithm can easily be modified, for example a crosswise change of perspective requires transposition of the 2D chromatogram matrix: $\mathbf{H}_i(j) = [\mathbf{H}_j(i)]^T$. All isometric projections in this book have been made by the algorithm described above.

The isometric plot is useful for qualitative screening of the 2D chromatogram and in getting an overall idea of the process under study. In chromatography, isometric plots have become popular with the appearance of rapid scanning multichannel photodiode array detectors [92,93]. These detectors give a two-dimensional output. The first dimension is wavelength in absorption spectra and the other is the chromatogram running time. However, it is difficult to perform any measurements from an isometric plot.

3.6.2 Slices of two-dimensional chromatograms

The slice of a two-dimensional chromatogram is a one-dimensional chromatogram recorded at a particular value of a second parameter. A set of 1D chromatograms obviously fully characterizes a 2D chromatogram. The intensity and location of the peaks on a 2D chromatogram could easily be measured in the direction of both dimensions. However, some effort is necessary to imagine the whole process, because direct vizualization is lost. Slices of two-dimensional chromatograms are shown in Fig. 5.22, p. 176.

3.6.3 Contour plots

A contour plot (a map) is a well known method of describing a surface in cartography. The same principle is directly applicable to the presentation of 2D chromatograms, which indeed conceptually resemble a mountain area.

The contour is obtained by cutting the 2D chromatogram surface by a plane that is parallel to the xy-plane and projecting the cut line onto the xy-plane. Since a 2D chromatogram is presented digitally it is determined only at the points $j\Delta x$, $i\Delta y$, so the whole area of a 2D chromatogram is divided into unit squares with an area $\Delta x \Delta y$, and the 2D chromatogram is determined only at the corners of these squares. To construct a contour line cutting a plane of the 2D chromatogram surface, the latter should be determined at the other points of the unit squares as well. The simplest way to do this is to approximate the 2D chromatogram surface between the unit corners with a bilinear surface. Let us consider a particular unit square where the 2D chromatogram surface has the values $H_j(i)$, $H_{j+1}(i)$, $H_j(i+1)$, $H_{j+1}(i+1)$. The bilinear surface can be constructed as follows.

(1) Straight lines are drawn between two point pairs $H_j(i)$, $H_{j+1}(i)$ and $H_j(i+1)$, $H_{j+1}(i+1)$.
() The bilinear surface is formed by a straight line that moves so that its ends lie on the straight lines described in (1). The line must move in such a way that the co-ordinates of its ends always stay equal to one another. Analytically the equation for the bilinear surface $z(x,y)$, determined over the unit square is given by [91]

$$z(x,y) = z_1(1-x)(1-y) + z_2(1-y) + z_3 y(1-x) + z_4 xy$$

where z_1, z_2, z_3 and z_4 are constants. A BASIC program listing for constructing a bilinear surface is given in Appendix 4.
(3) With a bilinear representation of one part of the 2D chromatogram the cutting line between the bilinear surface and the plane can be constructed by standard methods of analytical geometry.

A bilinear surface is shown in Fig. 3.14 and an example of a contour plot is shown in Fig. 5.25 (p. 180). All the contour plots in this book were made by using the algorithm above. The contour plot presentation of a data matrix such as the 2D chromatogram is more convenient, although perhaps less familiar, than the isometric plot. Contour plots are extensively applied to 2D NMR spectroscopy and also in chromatography [92,93]. In a contour plot all the information is displayed on a single plane, so small peaks are not hidden by large foreground ones as sometimes happens in an isometric presentation. However, the peaks with intensities below the cutting level are not represented in the contour plot at all.

One possible way of producing a contour plot is to track each contour around the

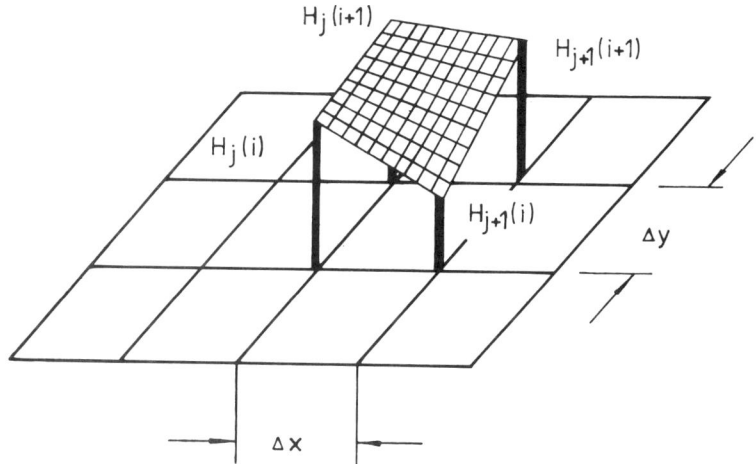

Fig. 3.14 — Bilinear surface.

xy-plane and to plot the 2D chromatogram contour by contour. This method is not suitable when a microcomputer is used for plotting contours [93]. The full data matrix cannot be held in the microcomputer memory and to draw a 2D chromatogram contour by contour requires a sequential transfer of each spectrum from the disk for each contour plotted. This is very time-consuming. Another solution [93], also used for presenting contours in this book, is to keep only two adjacent 1D chromatograms in the computer memory, simultaneously compute the bilinear surface between these chromatograms and draw only part of the corresponding contours. The whole picture can easily be drawn by considering all 1D chromatogram pairs one after another.

Contour plotting requires more computer time than does an isometric projection or drawing a set of 1D slices of a 2D chromatogram, because a large number of unit squares must be taken into account for a possible intersect between the plane and the surface. Using compilers of a high-level programming language (commonly BASIC in microcomputers) instead of an interpreter of a language considerably speeds up the computations. We have had good experience with the Applesoft Compiler (TASC) manufactured by Microsoft. These comments reflect the state of the art of microcomputers only. The power of the next generation of personal computers will probably be so high that much more elegant algorithms will be used for 2D data presentation.

From a contour plot the retention times of peaks are readily obtained. The peak shape may be deduced from the spacing between adjacent contours. These properties of the contour plot are useful for identification of unknown peaks by means of standards.

REFERENCES

[1] E. Bartoldi, *Bruker Report,* 1979, **2**, 2.
[2] R. N. Bracewell, *The Fourier Transform and its Application,* McGraw-Hill, New York, 1978.

[3] L. A. Rusinov, *Automization of Analytical Systems for Determination of Composition and Quality of Materials*, p. 22, Khimiya, Leningrad, 1984 (in Russian).
[4] P. C. Kelly and G. Horlick, *Anal. Chem.*, 1973, **45**, 518.
[5] Yu. G. Tkach, *Russian J. Phys. Chem.*, 1977, **51**, 1126.
[6] H. R. Murdock, Jr., *Anal. Chem.*, 1970, **42**, 687.
[7] P. J. Macnaughtan, Jr. and L. B. Rogers, *Anal. Chem.*, 1971, **43**, 822.
[8] E., Küllik, M. Kaljurand and M. Lamberg, *Russian J. Anal. Chem.*, 1986, **41**, 1425.
[9] M. J. E. Golay, in *75 Years of Chromatography — a Historical Dialogue*, L. S. Ettre and A. Zlatkis (eds.), p. 109. Elsevier, Amsterdam, 1979.
[10] D. H. Desty, in *Advances in Chromatography*, J. C. Giddings and R. A. Keller (eds.) Vol. 1, p. 199, Dekker, New York, 1965.
[11] G. Gaspar, P. Arpino, and G. Guiochon, *J. Chromatog Sci.*, 1977, **15**, 256.
[12] G. Gaspar, R. Annino, C. Vidal-Madjar and G. Guiochon, *Anal. Chem.*, 1978, **50**, 1512.
[13] G. Guiochon, *Anal. Chem.*, 1978, **50**, 1812.
[14] C. P. M. Schutjes, E. A. Vermeer, J. A. Rijks and C. A. Cramers, *J. Chromatog.*, 1982, **253**, 1.
[15] C. P. M. Schutjes, C. A. Cramers, C. Vidal-Madjar and G. Guiochon, *J. Chromatog.*, 1983, **279**, 269.
[16] R. Annino and J. Leone, *J. Chromatog. Sci.*, 1982, **20**, 19.
[17] R. J. Jonker, H. Poppe and J. F. K. Huber, *Anal. Chem.*, 1982, **54**, 2447.
[18] J. L. DiCesare, M. W. Dong and F. L. Vandemark, *Am. Lab.*, 1981, **13**, No. 8, 52.
[19] J. P. Gourlia and J. Bordet, *J. Chromatog. Sci.*, 1981, **19**, 35.
[20] J. L. DiCesare, M. W. Dong and L. S. Ettre, *Introduction to High-Speed Liquid Chromatography*, Perkin-Elmer, Norwalk, 1981.
[21] J. Sternberg, in *Advances in Chromatography*, J. C. Giddings and R. A. Keller (eds.), Vol. 2, p. 205. Dekker, New York, 1966.
[22] M. C. Harvey and S. D. Stearns, *Anal. Chem.*, 1984, **56**, 837.
[23] R. L. Wade and S. P. Cram, *Anal. Chem.*, 1972, **44**, 131.
[24] B. A. Ewels and R. D. Sacks, *Anal. Chem.*, 1985, **57**, 2774.
[25] D. R. Deans, *Chromatographia*, 1968, **1**, 18.
[26] K. Izawa, K. Furuta, T. Fujiwara and N. Suyama, *Ind. Chim. Belge*, 1967, **32**, 223.
[27] R. Annino and L. E. Bullock, *Anal. Chem.*, 1973, **45**, 1221.
[28] M. Kaljurand and E. Küllik, *J. Chromatog.*, 1979, **171**, 243.
[29] K. R. Godfrey and M. Devenish, *Meas. Contr.*, 1969, **2**, 228.
[30] R. Annino and L. E. Bullock, in *Gas Chromatography 1972*, S. G. Perry and E. R. Adlard (eds.), p. 171. Appl. Science Publishers, London, 1973.
[31] S. R. Frazer, *Information Extraction in Chromatography Using Correlation Technique*, Dissertation, Univ. of Arizona, 1985.
[32] M. Koel, M. Kaljurand and E. Küllik, *Proc. Estonian SSR Acad. Sci.*, 1983, **32**, 125.
[33] R. Hoffman, M. M. Gupta and P. N. Nikiforuk, *Proc. IEE*, 1977, **119**, 237.
[34] E. Küllik and M. Kaljurand, *Anal. Chim. Acta*, 1986, **181**, 51.
[35] M. Kaljurand and E. Küllik, *Trends Anal. Chem.*, 1985, **4**, 200.
[36] E. Küllik and M. Kaljurand, *Computer Chromatography with Multiple Input*, Preprint Estonian SSR Acad. Sci., Tallinn, 1986 (in Russian).
[37] S. Z. Roginskii, M. I. Yanovskii and A. D. Berman, *The Fundamentals of Chromatographic Application in Catalysis*, Nauka, Moscow, 1972 (in Russian).
[38] S. M. Langer and J. E. Patton, in *New Developments in Gas Chromatography*, H. Purnell (ed.), p. 293. Wiley-Interscience, New York, 1973.
[39] T. Paryjczak, *Gas Chromatography in Adsorption and Catalysis*, Horwood, Chichester, 1985.
[40] J. Coca and S. H. Langer, *ChemTech*, 1983, **13**, 682.
[41] L. R. Snyder, J. W. Dolan and Sj. van der Wal, *J. Chromatog.*, 1981, **203**, 3.
[42] L. R. Snyder, *U.S. Patent* 4204952, 27 May 1980.
[43] H. Boer and P. Van Arkel, *Chromatographia*, **4**, 300.
[44] H. Boer, in *75 Years of Chromatography — a Historical Dialogue*, L. Ettre and A. Zlatkis (eds.) p. 11. Elsevier, Amsterdam, 1979.
[45] J. F. K. Huber, R. van der Linden, E. Ecker and M. Oreans, *J. Chromatog.*, 1978, **83**, 267.
[46] J. F. K. Huber, in *75 Years of Chromatography — a Historical Dialogue*, L. Ettre and A. Zlatkis (eds.), p. 159. Elsevier, Amsterdam, 1979.
[47] D. R. Deans, *J. Chromatog.*, 1981, **203**, 19.
[48] C. J. Little and O. Stahel, *Int. Lab.*, 1984, **14**, June, 26.
[49] F. Müller and U. K. Goekeler, *Int. Lab.*, 1985, **15**, June, 42.
[50] *Chrompack News*, 1985, **12**, 1.
[51] D. Wright, *Int. Lab.*, 1985, **15**, Nov./Dec., 52.

References

[52] D. H. Freeman, *Anal. Chem.*, 1981, **53**, 2.
[53] G. C. Giddings, *Anal. Chem.*, 1984, **56**, 1258A.
[54] F. W. Willmott, I. Mackenzie and R. J. Dolphin, *J. Chromatog.*, 1978, **167**, 31.
[55] R. J. Dolphin, F. W. Willmott, A. D. Mills and L. P. J. Hoogveen, *J. Chromatog.*, 1976, **122**, 259.
[56] F. Erni and R. W. Frei, *J. Chromatog.*, 1978, **149**, 561.
[57] F. Erni, H. P. Keller, C. Morin and M. Schmitt, *J. Chromatog.*, 1981, **204**, 65.
[58] M. C. Harvey and S. D. Stearns, *Am. Lab.*, 1981, **13**, No. 2, 151.
[59] *Valco Valves for Gas- and Liquid Chromatography*, Chrompack, The Netherlands, 1978.
[60] D. R. Deans, *J. Chromatog.*, 1984, **289**, 43.
[61] H. Brötell, G. Rietz, S. Sandqvist, M. Berg and H. Ehrsson, *HRC&CC*, 1982, **5**, 596.
[62] G. Schomburg, F. Weeke, F. Müller and M. Oreans, *Chromatographia*, 1982, **16**, 87.
[63] K. Warman, *J. Anal. Appl. Pyrol.*, 1984, **7**, 137.
[64] R. J. Miller, S. D. Stearns and R. R. Freeman, *HRC&CC*, 1979, **2**, 55.
[65] J. C. M. Wessels and R. P. M. Dooper, *J. Chromatog.*, 1983, **279**, 349.
[66] G. Schomburg, H. Husmann and F. Weeke, *J. Chromatog.*, 1975, **112**, 205.
[67] G. Schomburg, *J. Chromatog. Sci.*, 1983, **21**, 97.
[68] F. Müller, *Int. Lab.*, 1983, **13**, July/Aug., 56.
[69] H.-J. Stan and D. Mrowetz, *J. Chromatog.*, 1983, **279**, 173.
[70] B. M. Gordon, C. E. Rix and M. R. Borquerding, *J. Chromatog. Sci.*, 1985, **23**, 1.
[71] W. Bertsch, *HRC&CC*, 1978, **1**, 85.
[72] W. Bertsch, *HRC&CC*, 1978, **1**, 187.
[73] W. Bertsch, *HRC&CC*, 1978, **1**, 289.
[74] *J. Chromatog. Sci.*, 1986, **24**, No. 1.
[75] C. Liteanu and I. Rîcă, *Statistical Theory and Methodology of Trace Analysis*, p. 96. Horwood, Chichester, 1980.
[76] J. C. Giddings, *Anal. Chem.*, 1967, **39**, 1027.
[77] J. C. Giddings, *Sepn. Sci.*, 1969, **4**, 181.
[78] E. Grushka, *Anal. Chem.*, 1970, **42**, 1142.
[79] J. Sevčik, *J. Chromatog.*, 1979, **186**, 129.
[80] R. J. Phillips, K. A. Knauss and R. R. Freeman, *HRC&CC*, 1982, **5**, 546.
[81] J. F. K. Huber and H. C. Smit, *Z. Anal. Chem.*, 1969, **245**, 84.
[82] J. F. K. Huber, E. Kenndler and G. Reich, *J. Chromatog.*, 1979, **172**, 15.
[83] J. V. Hindshaw and L. S. Ettre, *Chromatographia*, 1986, **21**, 561.
[84] R. E. Kaiser, R. I. Rieder, L. Leming, L. Blomberg and P. Kusz, *HRC&CC*, 1985, **8**, 580.
[85] R. E. Kaiser and R. I. Rieder, *5th Danube Symp. Chromatography*, Abstracts, p. 4. Nauka, Moscow, 1985.
[86] T. Toth, H. van Cruchten and J. Rijks, *6th Int. Symp. Capillary Chromatography*, P. Sandra (ed.), p. 769. Hüthig Verlag, Heidelberg, 1985.
[87] P. Sandra, F. David, M. Boot, O. Diricks, M. Verstappe and M. Verzele, *HRC&CC*, 1985, **8**, 782.
[88] D. R. van Hare and L. B. Rogers, *Anal. Chem.*, 1985, **57**, 628.
[89] S. N. Deming and S. L. Morgan, *Anal Chem.*, 1973, **45**, 278A.
[90] S. L. Morgan and S. N. Deming, *J. Chromatog.*, 1975, **112**, 267.
[91] D. F. Rogers and A. J. Adams, *Mathematical Elements for Computer Graphics*, McGraw-Hill, New York, 1976.
[92] A. F. Fell, H. P. Scott, R. Gill and A. C. Moffat, *J. Chromatog.*, 1983, **282**, 123.
[93] B. J. Clark, A. F. Fell, H. P. Scott and D. Westerlund, *J. Chromatog.*, 1984, **286**, 261.

4
Instrumentation

4.1 BASIC SET-UP OF MULTIPLE-INPUT COMPUTERIZED CHROMATOGRAPHY

The necessary hardware for multiple-input computerized chromatography is presented in Fig. 4.1. The system consists of four parts: a sample source, an input system, a chromatograph, and a computer.

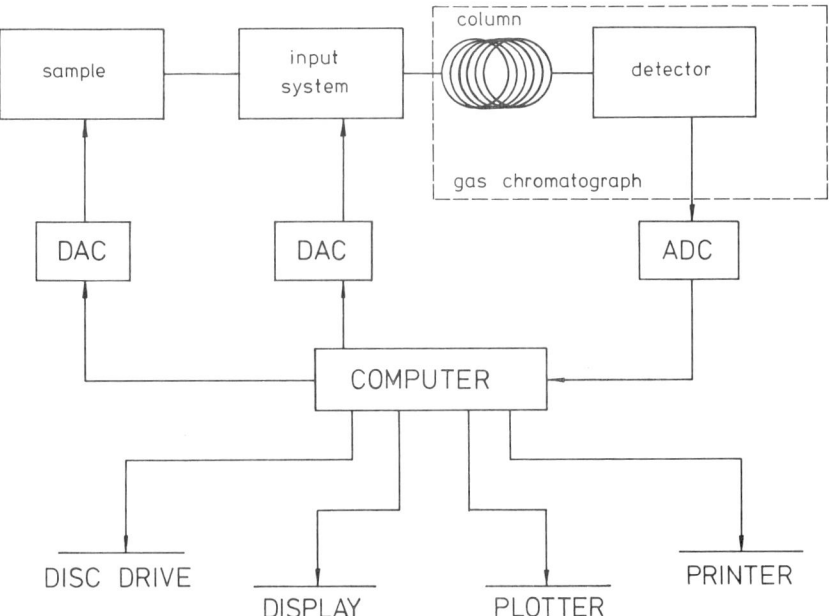

Fig. 4.1 — Basic scheme of a computerized chromatograph capable of multiple injections.

The sample source is an independent unit supplying a continuous sample flow to the input system. Its nature depends on the problem set. The sample source may be a diffusion call, a head-space sampler, a chemical reactor, or a simple flask with a sample solution.

The input system is the most critical part of the equipment. It is the unit that transforms the electrical signals generated by the computer into the necessary concentration changes. The performance of the whole system is determined by that of the input. The input must be reliable and ensure that the sample concentration at the column inlet is that produced by the electrical wave-form pattern generated by the computer. Unfortunately, no such ideal input system exists. In liquid chromatography mechanical valves are used to modulate the input whilst in gas chromatography, mechanical valves and pneumatic switches are in use as modulators. The latest studies by Phillips et al. [1] have shown that a direct modulation of the input flow by various physical and chemical means is also possible.

Any chromatograph may be adapted to work in the multiple input mode. If the purpose of the experiment is column switching, then the input system and chromatograph in Fig. 4.1 should be replaced by several input systems and columns in series or parallel. The use of detectors with non-linear response has been restricted in correlation chromatography. Column temperature programming is difficult or impractical in some applications of the multiple input system (e.g. in correlation chromatography).

A digital-to-analogue converter (DAC) converts the sequence of numbers generated by the computer, into a continuous electrical signal (voltage), and an analogue-to-digital converter (ADC) reverses the process.

The computer is usually microprocessor-based and has several programs which control the input system and log the data. The peripheral equipment (keyboard, disc drive, display, plotter, and printer) enables the chromatographer to control the computer, and most personal computers have these peripherals. Below, we shall describe in more detail each part of the computerized multiple injection chromatograph.

4.2 SAMPLE SOURCE

Though a single injection requires only a few μl of sample, a complex input requires much more. Except for some rare cases of biological samples, the required amount of sample is usually readily available. Moreover, several applications of correlation chromatography are particularly suited to studies of large amounts of dilute samples (e.g. air and water). Below, we consider six possible sample sources. Three of these are for calibration purposes: an exponential dilution flask, a diffusion tube, and a permeation tube. The other three are for use in the analysis. Two of them are very simple: environmental air, and a vessel containing a liquid sample. The third is more sophisticated — a chemical microreactor.

4.2.1 Individual sample sources

The simplest sample source is a container filled with the sample that is to be pumped through the liquid chromatograph [2]. A corresponding sample source in gas chromatography is environmental air forced to flow through the input system by an aspirator; this has been used to determine methane in environmental air [3].

In gas chromatography, there are several types of sample source. Some of these supply a continuous sample flow and are thus of interest in multiple injection chromatography.

An exponential dilution flask (Fig. 4.2) is a vessel with a carrier gas inlet, a sample

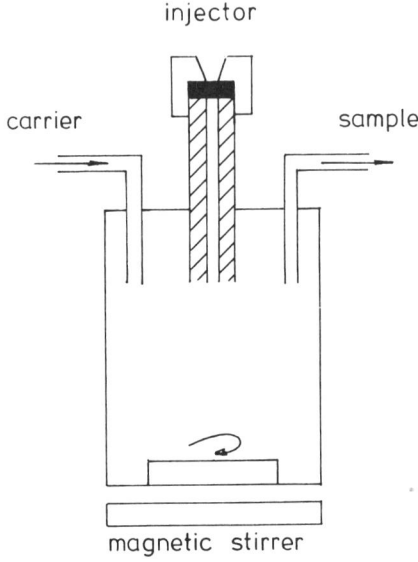

Fig. 4.2 — Exponential dilution flask.

output, and an ordinary GC injector. If a known amount of gas sample is injected into the vessel and the gas mixture is stirred vigorously by a magnetic stirrer, the sample concentration decreases exponentially with time according to the following formula:

$$C(t) = C_0 \exp(-Ft/V) \tag{4.1}$$

where $C(t)$ is the sample concentration at time t, C_0 is the initial concentration, F is the volumetric carrier gas flow-rate and V is the vessel volume. Equation (4.1) can easily be derived by taking into account that the amount of sample dm leaving the vessel during an infinitesimally short period of time dt is equal to $-CF\,dt$. The exponential dilution flask was first used by Lovelock [4]. It has also been used to calibrate the correlation chromatograph in methane determination [5]. Because the output flow is non-stationary, a certain amount of correlation noise appears in the computed chromatograms. However, the level of this noise can be kept low by controlling the vessel volume V and especially by controlling the carrier gas flow-rate F. To estimate the signal to (correlation) noise ratio, we can use Eq. (2.22): $s_{\Delta H} = \gamma b \sqrt{n}\, \Delta t$. Here b is the amount of change in the input function during the sampling interval and γ is a dimensionless coefficient; in the case of exponential dilution of the sample differentiation of Eq. (4.1) gives $b = -C(t)F/V$, and the signal to (correlation) noise ratio (S/N) is

$$\frac{S}{N} = \frac{C(t)}{s_{\Delta H}} = \frac{V}{\gamma\sqrt{n}\,F\Delta t} \qquad (4.2)$$

Hence, according to Eq. (4.2), increasing the PRBS length decreases the signal to noise ratio, but using a suitable V/F ratio enables the use of exponential dilution input for the calibration of the correlation chromatograph without any problems with correlation noise. If we take $V = 500$ ml, $n = 1024$, $\Delta t = 1$ sec, $F = 10$ ml/min and $\gamma = 1$, we obtain from Eq. (4.2) $S/N = 94$, i.e. the correlation noise can be neglected. This result can also be obtained in another way: the concentration change during the experiment is approximately 30% and because of the multiplex advantage this variation is suppressed by a factor of $\sqrt{1024}/2 = 16$, which leads to a correlation noise level approximately two orders of magnitude lower than the signal level.

The diffusion cell is a very simple sample generation device, which is why it is widely used (Fig. 4.3). Although theoretically the output concentration is non-

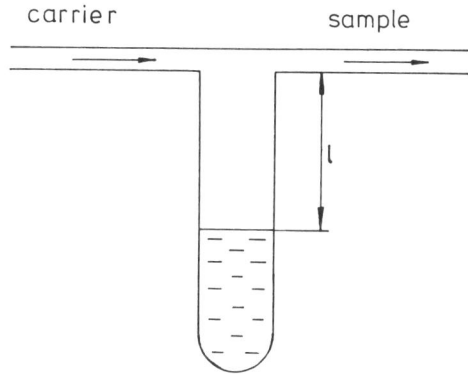

Fig. 4.3 — Diffusion cell: l — sample diffusion distance to the carrier gas flow.

stationary, in the case of long experimental times the variation in concentration is very small. The rate of sample diffusion into the carrier gas is given by [6]:

$$r = \frac{DAMP}{RTl}\ln\left(\frac{P}{P-p}\right) \qquad (4.3)$$

where r is the rate of diffusion of vapour out of the diffusion cell (g/sec), D is the molecular diffusion coefficient of the vapour (cm^2/sec), M is the molecular weight of the vapour, P is the total pressure in the diffusion cell (atm), p is the partial pressure (atm) of the vapour at the absolute temperature T (K) of the cell, A is the cross-sectional area of the diffusion cell (cm^2), R is the gas constant (l. atm. mole^{-1}), and l is the diffusional path-length (cm). The rate of sample evolution is determined by the temperature and the geometrical dimensions of the cell. Whilst the exponential dilution flask is suitable for calibration with substances which are gases at ambient temperature, the diffusion cell is very convenient for generating sample flows when the sample is a liquid at ambient or near-ambient temperature. The amount of sample evolved can easily be changed by altering the cell temperature. This cell can be made very simply by incorporating a test-tube of suitable dimensions in the conventional injection port of a chromatograph.

From Eq. (4.3), the diffusion rate is inversely proportional to the cell length. It is of interest to express the diffusion rate as a function of experimental time. The amount of sample dm in the cell volume dV is ρAdl, where ρ is the sample density. Since $r=$ dm/dt, we obtain

$$\frac{\rho A \mathrm{d}l}{\mathrm{d}t} = \frac{DAMP}{RTl} \ln \frac{P}{P-p} \qquad (4.4)$$

Integrating Eq. (4.4) gives $l \sim \sqrt{t}$, i.e. the rate of sample evolution from the diffusion cell decreases as the inverse of the square root of the duration of the experiment. If the carrier gas flow through the cell is fast and the cell volume is small, then all the sample that diffuses into the carrier is removed from the cell and the sample concentration, $C(t)$, in the carrier will be r/F (i.e. during 1 sec r mass units of sample diffuse into F volume units of carrier). Hence, the concentration also decreases as the inverse of the square root of the experimental time, $C(t) \sim t^{-1/2}$. The derivative of the concentration with respect to time is $C'(t) \sim t^{-3/2}$. From Eq. (4.2), the signal to (correlation) noise ratio is proportional to $C(t)/C'(t)$ and it follows that in the case of the diffusion cell this ratio improves linearly with time, i.e. $S/N \sim t$.

If the cell temperature and carrier gas flow-rate are carefully controlled, the diffusion cell is a good sample source for multiple input chromatography. After an initial approach to the steady state the cell gives an almost stationary output (variations are less than a few per cent). The steady-state condition (within 1%) will be reached at $t > l^2/2D$ [6]. A description of more sophisticated diffusion cells can be found in [7].

Another sample source for the calibration of correlation gas chromatographs is a permeation tube [8]. This is a simple device in which a piece of polymer tubing is filled with a certain amount of substance which diffuses through the tube wall into the surrounding flow. The permeation tube supplies a constant concentration flow over a period of several months. The sample concentrations that can be generated are generally lower then those obtained with diffusion tubes.

Chemical microreactors are also widely used as sample sources. In chromatography, three different reactor types are of interest: the periodic reactor, the impulse reactor, and the continuous-flow reactor [9]. In the periodic reactor, the reaction is performed in a conventional reaction vessel kept at constant temperature, and filled with one reagent and (or) a catalyst. The reaction is initiated by injecting the other reagent(s). The course of the reaction is studied by taking samples at equal intervals of time and injecting them into the chromatograph.

The impulse reactor is commonly a glass or metal tube filled with a catalyst. The reactor is continuously flushed with the carrier gas. The reaction is initiated by injecting the reagent into the carrier gas flow and the products are carried from the reactor to the input system of the chromatograph. In some experiments the input system and also the analytical column may be absent. Products and reagents are simultaneously measured by the detector. This is called on-column reaction chromatography. The impulse reactor is also called a 'chromatographic reactor' because two different processes take place: a chemical reaction and the chromatographic separation of the products and reagents [10]. Owing to the separation of products and reagents an interesting phenomenon takes place in this reactor: the product yields are generally higher than would be expected on thermodynamic grounds, because the equilibrium is shifted in the direction of increasing conversion. The output signal from this reactor has a complicated pattern of overlapping reagent and product peaks. In the simplest case of the first-order reaction, $A \to B$, the output signal is an exponentially decreasing peak of B superimposed on the Gaussian peak of A (Fig. 1.3, p. 20). As already suggested in Section 3.2, two-dimensional methods should give interesting results in studying this type of reaction.

In the continuous-flow reactor, the reagent is continuously fed to the reactor, which is a tube filled with a catalyst, and the products are directed to the input system. Because of a continuous feed of reagents, a steady-state distribution of concentrations is established over the catalyst. This kind of microreactor is similar in operation to larger scale industrial reactors. The only variations in reactor output; when the temperature and pressure are constant, are due to changes in the catalyst activity.

Special types of chemical microreactors, [e.g. pyrolytic and thermal degradation (desorption) reactors], will be considered more thoroughly in the next section. They are widely used in analytical and applied pyrolysis where the characterization of the properties of solid compounds (e.g. polymers) is of interest.

4.2.2 Pyrolysis reactors

Pyrolysis reactors can be classified according to the mode and nature of the heat source. Thus there are two major groups of pyrolysis systems. Continuous pyrolysis systems include a microreactor consisting of a tubular furnace or a resistively heated filament. The reactor normally has an oven with low dead-volume, a uniformly heated zone, and a sample holder for inserting the sample into the degradation zone and removing it again. It is common for the sample to be inserted into a small silica tube placed in the hot zone.

With pulse-mode pyrolysers the energy source may be a resistively heated filament, or an inductively heated ferromagnetic metal wire or filament. The inductively heated pyrolysers are known as Curie-point pyrolysers. Other degradation systems, such as electric discharge, laser pyrolysis or radiation-induced pyrolysis can also be regarded as pulse-mode systems.

There is another classification of pyrolysis systems based on the heat source of the reactor. There are three basic groups: (1) filament-heated pyrolysers, (2) Curie-point pyrolysers and (3) furnace pyrolysers. The classification is summarized in Table 4.1.

Table 4.1 — Classification of pyrolysis reactors

Heat source	Mode of pyrolysis	
	Pulse	Continuous
Resistive heating	filament	filament
Inductive heating	Curie-point wire	Curie-point wire
Conductive heating	—	furnace tube
Electric discharge	arc	—
Laser impulse	microreactor with window	—

The operating principle of filament-heated pyrolysers is that an electric current is passed through a resistance wire where the power is dissipated and increases the temperature of the conductor. This type of reactor was widely used in the early days of pyrolysis gas chromatography, because of its simple construction. Its disadvantage was the relatively long temperature rise-time, of the order of 10–30 sec in a fixed voltage system, and the poor reproducibility of the pyrolysis temperature.

In the Curie-point pyrolyser, the heating element is a ferromagnetic wire inductively heated in a radiofrequency field to its Curie-point, i.e. the temperature at which the alloy becomes paramagnetic and ceases to absorb energy.

A furnace pyrolyser is a continuously and externally heated chamber into which the sample is introduced in a silica tube so that it is not in contact with the furnace wall. Theoretically, thermal equilibrium of the gaseous surroundings will eventually be attained at the pyrolysis temperature, which may be different from that of the furnace walls. There is some risk that the degradation products may migrate from the sample to another region of the heating chamber. This may lead to additional fragmentation and unwanted secondary reactions. The final product is normally a complex mixture of degradation products and their recombinations.

In comparison with the other techniques, laser pyrolysis is more specialized. The use of a laser beam to effect pyrolysis has several advantages. The degradation of the material takes place in a small volume (0.01 cm^3) and there is no need to grind the material to powder. The coherent energy beam allows selection of an area for analysis. The temperature rise, up to 3200°C in under 1 msec, is very rapid, compared with that of all the other degradation methods and cooling is also very rapid. When the energy pulse has decayed, the system returns to ambient temperature almost as fast as it has risen. The use of laser pyrolysis is complicated by difficulties in controlling the final temperature, but it is unique because very high energies may be dissipated very rapidly. Laser pyrolysers are used, for example, in organic geochemistry.

4.2.2.1 Sample preparation for pyrolysis reactors
Although the sample preparation time does not directly affect the final results of analysis, the preparation procedure is often important for obtaining reliable results.

The particular procedure used depends on the type of pyrolyser used. The pulse-mode heated-filament and Curie-point pyrolysers are the most popular, but the sample preparation needed is relatively time-consuming. Normally, if the polymer is soluble, a small part of the wire or ribbon is coated with a thin film of sample by application of a solution and evaporation of the solvent. Because it is in good contact with the heating element, the sample effectively follows the temperature-rise profile of the wire or ribbon.

However, there are many synthetic and natural polymers that are not soluble in chemically inert solvents. Generally, such samples can be ground or milled to a powder, which is then dispersed in a solvent such as methanol or carbon disulphide by ultrasonics, and the suspension is used to coat the wire. Materials such as rubber require to be frozen at very low temperature before milling.

Sample preparation involves several time-consuming steps. Moreover, after each pyrolysis the heating element should be cleaned in a flame or by heating the system before the next sample is placed on it and examined. The time for a particular operation in sample preparation varies widely; some typical times are given in Table 4.2. The data presented are based on personal experience and several literature sources. The preparation time can be shortened if several samples can be prepared as a batch.

Table 4.2 — Sample preparation time for analysis with Curie-point and filament type pyrolysers; mean values from several experiments

Operation	Time (min)	
	without extraction	with extraction
Wire or filament cleaning	4	4
Extraction with methanol	—	480–720
Sample grinding	10	10
Dissolving	10	—
Deposition of the sample	3	2
Drying in air	10	—
Evacuation of residual solvent in vacuum	20	30
Inserting sample into cell	3	3
Total time for preparation	60	529–769

4.2.2.2 *Temperature control in pyrolysis reactors*

The inner temperature of pyrolysis microreactors seems to be the most important parameter in determining the reproducibility of pyrolysis gas chromatography (PyGC). A great many home-made and a limited number of commercially available pyrolysers have been used to generate countless pyrolysis gas chromatograms. Satisfactory results have been obtained in laboratories where the analysis is performed by the same person with the same equipment. In most cases, the actual pyrolysis temperature remains uncertain. Both the difference in pyrolysis cell construction and the interlaboratory non-reproducibility of pyrolysis temperature are the main reasons why the huge number of published chromatograms of degradation products have only a limited application, and why in most cases, a personal atlas of pyrolysis chromatograms should be available.

An attempt to apply the Arrhenius equation to the degradation process gave somewhat unexpected results and was indicative of the influence of the temperature on the reproducibility of the amount of pyrolysis product. This can easily be demonstrated in the case of the low-temperature pyrolysis of polymers [11]. With the Arrhenius equation, the degradation process can be described by

$$dg(t)/dt = Z \exp(-E/RT)[g_0 - g(t)] \quad (4.5)$$

where $g(t)$ is the amount of the gas evolved during degradation, t is the time, g_0 is the total amount of the gas evolved, R is the gas constant, T is the absolute temperature, and Z and E are kinetic parameters characterizing the sample (pre-exponential factor and activation energy). Although Z and E cannot be assigned to definite physical or chemical parameters of a polymer, Eq. (4.5) fits the evolved gas curve, $g(t)$, quite well and the Z and E values are available from thermogravimetric measurements for a huge number of polymers. This justifies the application of Eq. (4.5) in the present study.

In a degradation experiment, samples are taken from the gas stream flowing through the reactor. The amount of sample, Δg, directed to the chromatographic column during the time Δt, can be obtained, by modification of Eq. (4.5), as

$$\Delta g = g_0 Z \exp(-E/RT) \Delta t \quad (4.6)$$

If low-temperature pyrolysis is considered, it can be assumed that the polymer has been only slightly degraded, which means that $g(t) = 0$. It follows from Eq. (4.6) that the signal value depends linearly on the amount of sample, g_0, and on the sampling time Δt.

The influence of temperature is obtained by differentiating Eq. (4.6) with respect to T:

$$\delta(\Delta g)/\Delta g = (E/RT)(\delta T/T) \quad (4.7)$$

It follows that small variations in temperature will be amplified by the factor E/RT. For some thermo-stable polymers, at the temperatures at which degradation starts, E/RT can be as high as 50, which could lead to considerable errors. In fact, Eq. (4.7) explains quite clearly why interlaboratory reproducibility in pyrolysis gas chromatography is so difficult to achieve. Small variations in temperature produce great changes in the amount (and probably, in the composition) of the evolved gases.

The sample temperature behaviour during pyrolysis is also important. The temperature rise-time is a significant part of the overall pyrolysis time. The rate of heating determines the results of pyrolysis at the temperature chosen and the quantitative composition of the pyrolysis products from the same substance can vary from analysis to analysis, depending on the heating profile. In general, there are two possible descriptions of the temperature profile in a sample during the heating time. The ideal temperature profile (Fig. 4.4A) is that from instantaneous heating of the sample up to the equilibrium temperature (T_{eq}), holding it for some definite time and then infinitely rapid cooling to the initial temperature (T_i). Pyrolysis should be considered ideal if the sample is not degraded significantly either before the equilibrium temperature is reached or during the cooling down to the initial temperature.

The actual pyrolysis temperature profile (Fig.4.4B) differs from the ideal in the

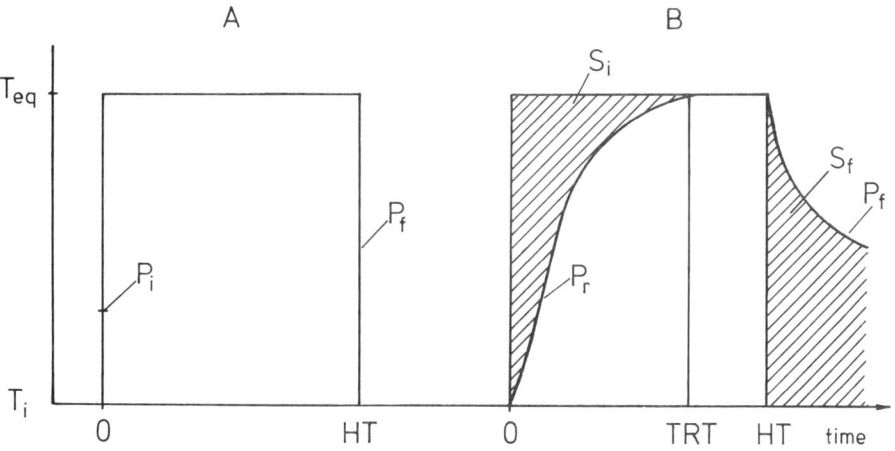

Fig. 4.4 — Ideal (A) and real (B) temperature–time profiles: TRT — temperature rise-time; HT — heating time; T_i and T_{eq} — initial and equilibrium temperatures. P_i, P_r and P_f — ideal, real and final temperature profiles. S_i and S_f — areas characterizing differences between ideal and real temperature profiles.

initial (P_i) and final (P_f) parts. Although the heating time is equal in both the ideal and real cases the pyrolysis in the real case proceeds in a way determined by the operating parameters and pyrolyser construction. The temperature rise-time (TRT) to the equilibrium temperature is a finite part of the heating time (HT). The areas S_i and S_f predetermine the course of the pyrolysis and are different for each pyrolyser. In conventional single-input pyrolysis gas chromatography, it has commonly been assumed that after the end of the heating the temperature inside the cell falls to the initial temperature in a cooling time which is much shorter than the time interval for separating the evolved gases in the column. This is certainly not true. However, the initial TRT profile affects the degradation results more than the cooling profile does.

Let us consider some experimental results of TRT measurements in filament type pyrolysers. Appreciation of the important role of temperature reproducibility in polymer degradation obviously directed Lehre and Robb's attention to the need to investigate the value of the actual pyrolysis temperature in more detail [12]. The first results of studies on pyrolysis temperature profiles were remarkable. Different values of heating current gave different temperature profiles during the same heating time (Fig. 4.5). The temperature rise profiles show that the maximum steady temperature is not reached immediately, but the temperature decay is sufficiently rapid. The rise-time to the equilibrium temperature is different for different current values and the profiles are far from having the ideal square shape. To improve the temperature–time profile, a booster current was used during the first second of the heating cycle. Oscilloscope traces showed the changes in the temperature profiles (Fig. 4.6). These profiles allow decisions to be made about the standardization of the temperature rise at the beginning of pyrolysis.

The effect of temperature rise time on the content of the evolved gases was clearly demonstrated with the isoprene–styrene copolymer sample, which was heated at different rates. Different heating rates produced different recoveries of the isoprene component. The amount of isoprene evolved increased with increasing heating rate, (Fig. 4.7) [13].

Subsequent measurements of TRTs confirmed the results of Lehre and Robb. As may be seen from Fig. 4.8, the times for the sample to reach the pyrolysis equilibrium temperature differs greatly for different types of pyrolysis unit. This is why care must be taken when comparing programs from different sources.

The temperature rise-time and actual pyrolysis time have also been carefully studied by Levy et al. [14]. The temperature–time profiles in a Curie-point and a filament pyrolyser were measured in different ways. Two methods were used to deposit the sample on the filament or ferromagnetic wire. A solution of polystyrene in benzene was placed on the filament at the point of contact between filament and thermocouple, and the solvent evaporated, by exposure to ambient air for 10 min, then at a pressure of 1 mmHg for 20 min to remove the residual solvent. The other sample, a powdered acrylic acid acrylonitrile copolymer, was sprinkled on the filament and gently heated to about 100°C. The pyrolysis temperature was measured by a photodiode and a chromel–alumel thermocouple (0.0075 cm in diameter). The thermocouple measured the temperature–time profile at the actual point of contact between the sample and filament. The results of the filament temperature measurements were, to some extent, unexpected. The oscilloscopic traces of the temperature–time profile of the filament without a sample (a) and with a sample (b) are quite different (Fig. 4.9). The area between the curves a and b is indicative of the sample lifetime on the filament. That kind of measurement permits the direct determination of the actual pyrolysis temperature, defined as that temperature at which the rate of energy consumed by the sample is equal to the whole power supplied to the system [14].

The inductive heating pyrolysis technique provides a repeatable pyrolyser temperature since the Curie-point temperature of a ferromagnetic material depends only on the composition of the alloy. In most cases, an Fe–Ni–Co alloy is used, and occasionally a Cr–Mo alloy is used. Table 4.3 lists some of the alloys used, and their composition. The compositions of the alloys permit choice of pyrolysis temperature in the range from 358°C (pure Ni) to 980°C (50% Fe–50% Co). The drawback of the Curie-point pyrolyser is that although it is possible to prepare a set of alloys which cover the whole temperature range 358–1128°C, each alloy gives only one discrete temperature (Fig. 4.10), so the method cannot be used for non-isothermal kinetic studies, in which the temperature is varied continuously.

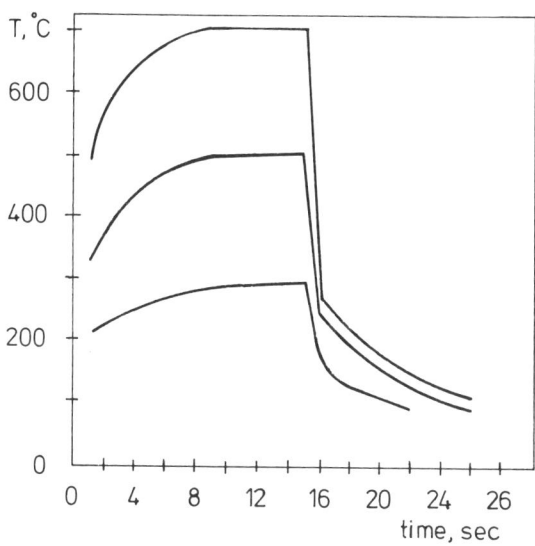

Fig. 4.5 — Different heating currents give different temperature profiles. Pre- and after-effect observed for a filament nominally at 300, 500 and 700°C for 15 sec. The warming up effect (pre-effect) is particularly serious, but the decay (after-effect) is rather rapid. (Adapted from [12]).

It is interesting that two different Fe–Ni alloys give the same Curie-point temperature, 510°C.

The composition of the ferromagnetic alloys affects not only the final temperature but also the temperature rise-time. It is also known that the same alloys give different temperature rise-times at different radiofrequency inputs (Fig. 4.11) [15].

Filament, Curie-point and furnace pyrolysers can be divided into two groups according to the temperature gradient in the reactor. In the filament and Curie-point mode (the pulse-mode) the sample is in direct contact with the heating element, which is normally placed in the middle of the pyrolysis unit (Fig. 4.12A). During heating, the degradation products rapidly move away from the hot zone. Secondary reactions are probably avoided or minimized in number because the carrier gas stream carries the products out of the hot zone to the cooler pyrolysis unit region and the connecting channel to the chromatograph. The temperature profile inside the reactor is symmetrical and the temperature at the walls of the reactor is lower than in the centre. Substances with high boiling points may condense on a cool wall surface, and the pyrolysis cell may be preheated to a temperature that would prevent this, but without influencing the sample composition.

In furnace pyrolysis the heating elements are placed uniformly around the reactor and the temperature of the externally heated wall of the pyrolysis cell is higher than that of the sample at the moment the sample reaches the heating zone (Fig. 4.12B). The primary pyrolysis products are released from the surface of the sample and expand into a hotter zone which is near the wall. The temperature profile is opposite to that of the filament wire or Curie-point pyrolysis system. The four symmetrically

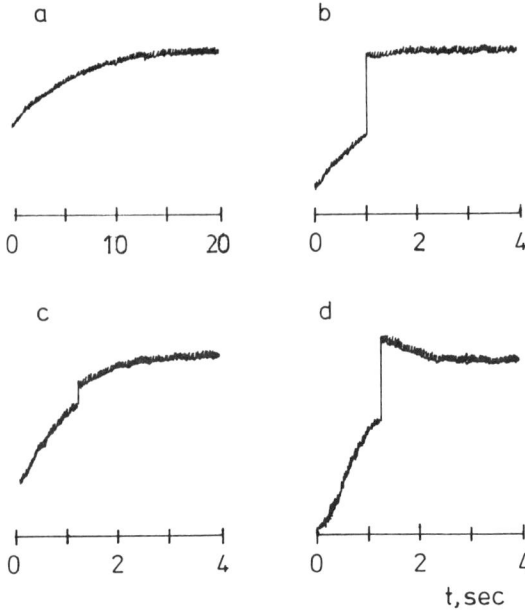

Fig. 4.6 — Oscillograms of temperature profiles after an initial boost current for 1 sec: a, no boost; b, correct boost; c, insufficient boost; d, excessive boost. (Adapted from [12]).

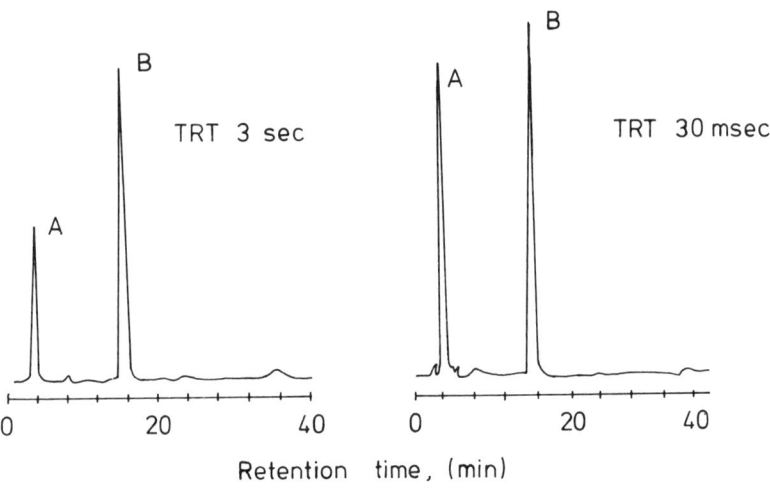

Fig. 4.7 — Pyrograms of an isoprene–styrene copolymer. The effect of temperature rise-time (TRT). Peaks: A, isoprene; B, sytrene. (Adapted from [13]).

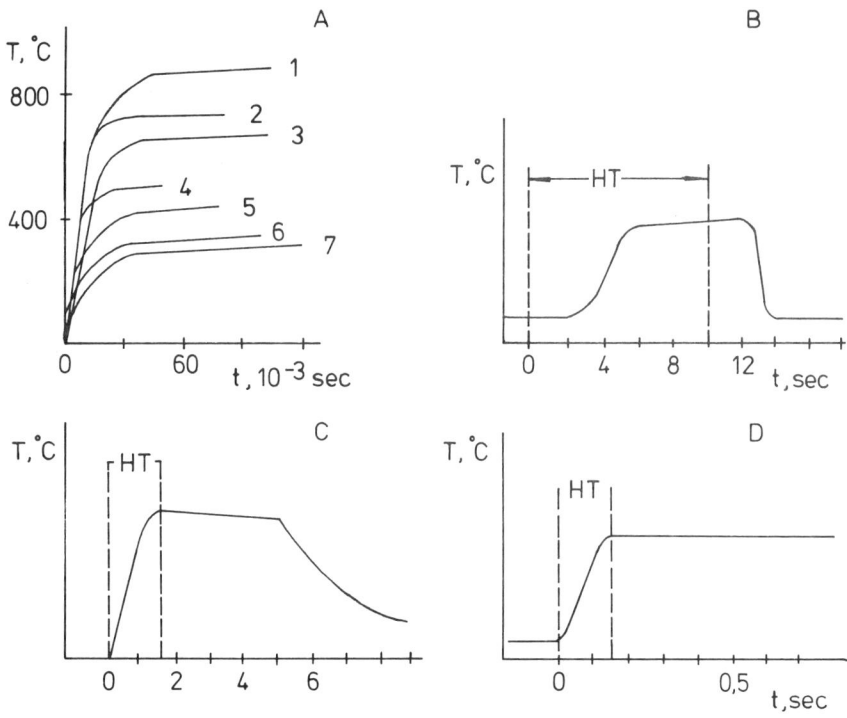

Fig. 4.8 — Temperature profiles dependent on pyrolysis reactor construction, compiled from several sources. (A) Curie-point pyrolyser: 1, CoNi (60:40); 2, Fe; 3, CoNi (33:67); 4, NiFe (60:40); 5, NiCrFe (51:1:48); 6, NiFe (45:55); 7, Ni; oscillator frequency 0.45 MHz. (B), Filament-type reactor directly heated by electric current from constant voltage source, heating time (HT) 10 sec. (C), Curie-point reactor for wire or filament, wire 0.5 mm in diameter, 30 W oscillator, HT 1.3 sec. (D) as for (C) but 2.5 kW oscillator, HT 120 msec.

placed heating cartridges (elements) give a different temperature profile (Fig. 4.13A). Sometimes the sample is placed in a small tube that rests on the bottom of the pyrolysis reactor (Fig. 4.13B). In this case, the temperature profile is not symmetrical. The container tube allows weighing of the pyrolysis residues. All the temperature profiles presented in Fig. 4.12 and Fig. 4.13 are schematic and are not based on real calculations.

Data in the literature show that attention has been paid to determining the actual pyrolysis temperature, but no attempt has been made to calculate a heat balance in a pyrolysis reactor, although the necessary data are available. A consideration of the thermal conductivity, heat capacity and heat insulation, could lead to a better understanding of the process and provide more detailed information about the real pyrolysis pattern and perhaps avoid exaggerations in future studies.

The so-called 'roast effect', in which the sample surface is at a higher temperature than the bulk of the sample has also been studied. Obviously this effect might be observed if the sample is large or the pyrolysis time is short. When thick or bulky

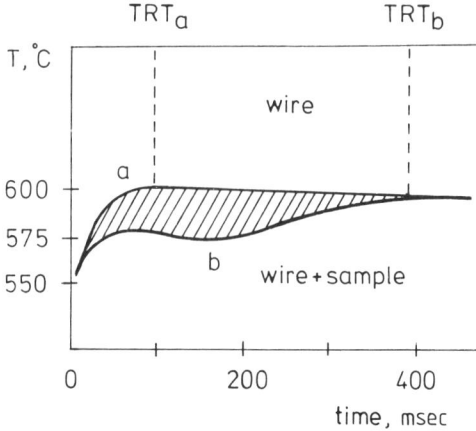

Fig. 4.9 — Temperature-time profiles of filament with and without sample, showing the true pyrolysis temperature. [Reprinted (adapted) with permission from R. L. Levy, D. L. Fanter and C. J. Wolf, *Anal. Chem.*, 1972, **44**, 38. Copyright 1972, American Chemical Society].

Table 4.3 — Curie points of some ferromagnetic metals

Curie point (°C)	Alloy composition (%)				
	Fe	Ni	Co	Cr	Mo
358	—	100	—	—	—
400	61.7	—	38.3	—	—
400	55	45	—	—	—
420	48	51	—	1	—
420	17	79	—	—	4
440	48	51	—	1	—
440	55	45	—	—	—
480	48	51	—	1	—
510	50.6	49.4	—	—	—
510	49	51	—	—	—
590	40	60	—	—	—
600	42	41	16	—	—
610	30	70	—	—	—
660	—	67	33	—	—
700	33	33	33	—	—
700	—	67	33	—	—
770	100	—	—	—	—
800	—	55	45	—	—
900	—	40	60	—	—
980	50	—	50	—	—
1128	—	—	100	—	—

samples are used, several effects can be present which are specific to pyrolysis gas chromatography: a temperature gradient exists between the interior and exterior of the sample and a uniform reaction temperature cannot be obtained; and products

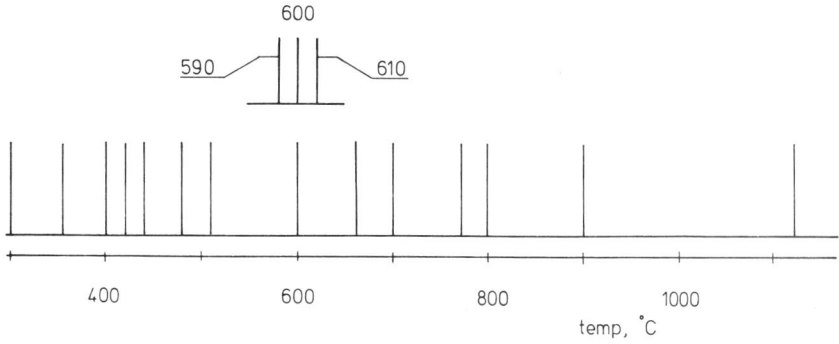

Fig. 4.10 — Pyrolysis temperatures available by using Curie-point pyrolysis technique. Random distribution of temperatures does not allow systematic study of the degradation process.

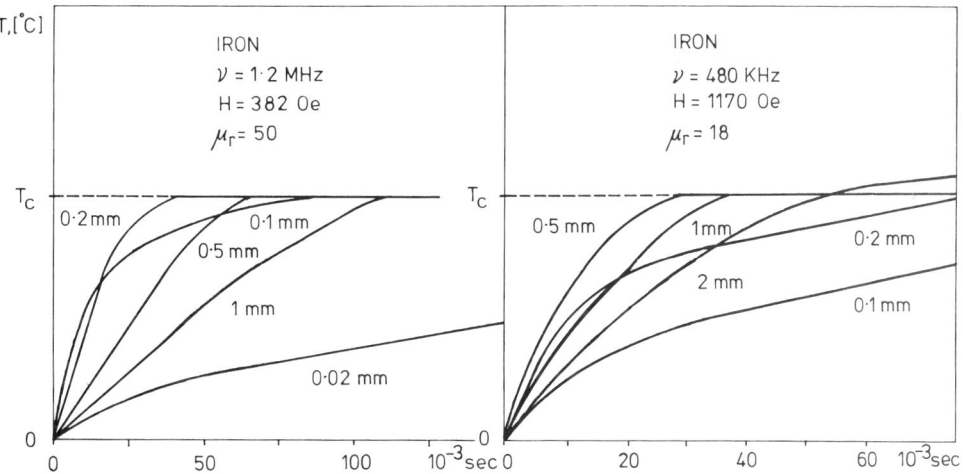

Fig. 4.11 — Temperature–time profiles of iron conductors for Curie-point pyrolysis for different operating parameters and wire diameters. (Adapted from [15]).

formed in the interior of the sample must diffuse to the surface for final volatilization. These problems arise with single-input pyrolysis gas chromatography. Accuracy and reproducibility are easier to achieve in the same laboratory but not between different laboratories. In any case, in pyrolysis gas chromatography reproducibility is still a major problem. A good account of pyrolysis reactors and the problems associated with them is given in Irwin's book [16].

Analytical non-reproducibility caused by temperature fluctuations could be

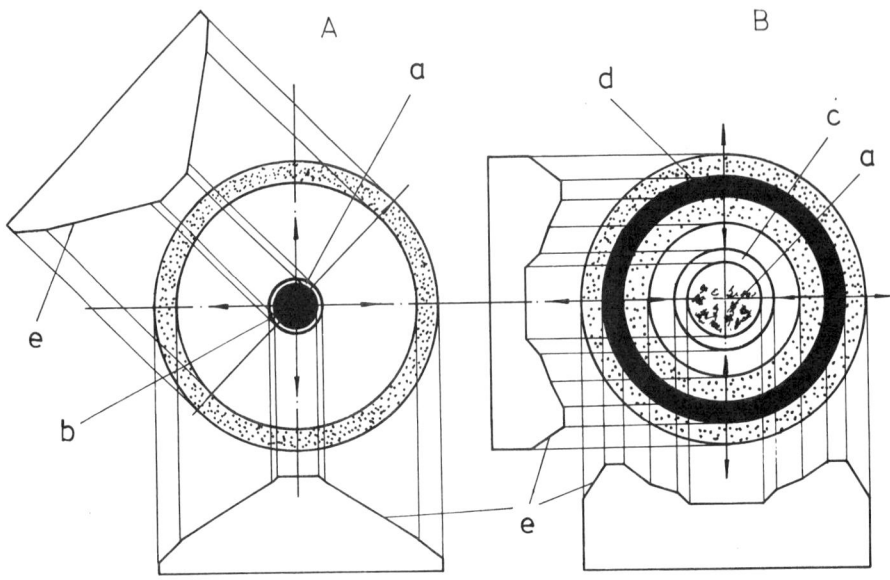

Fig. 4.12 — Distribution of temperature inside the reactor. (A), Curie-point type pyrolyser, sample placed round the wire. (B) Oven-type pyrolyser. a, Sample; b, Curie-point wire; c, silica container; d, heater; e, temperature profile.

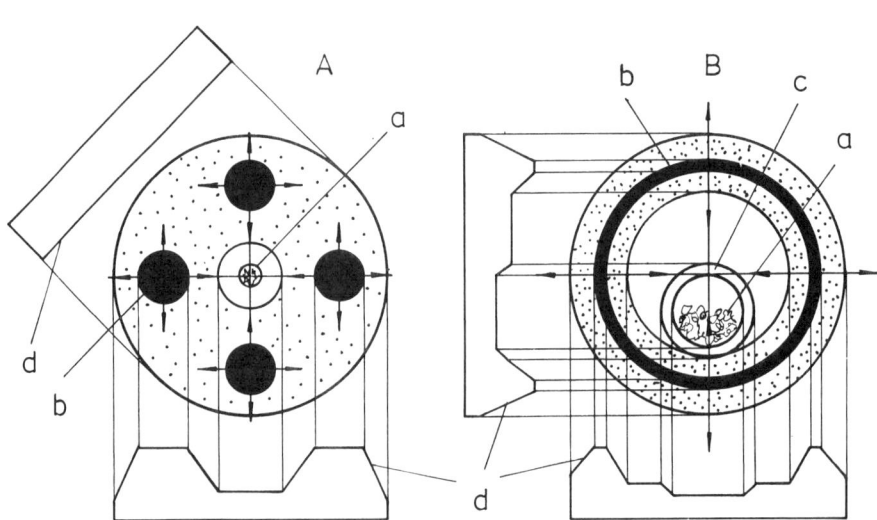

Fig. 4.13 — Distribution of temperature inside the reactor. (A) Non-uniform heating; (B) non-symmetrical placement of sample. a, Sample; b, heater; c, silica container; d, temperature profile.

eliminated by using only one sample, heating it over the whole temperature range of interest, and collecting samples of the evolved gases with a frequency allowing description of the degradation process with the required accuracy.

Let us assume that the temperature increases linearly. Then the temperature at some arbitrary point outside the sample is $T = T_0 + bt$ where T_0 is the initial temperature of the reactor, b is the heating rate (deg/sec) and t is time.

The basic equation describing the dependence of temperature at any point of the sample on time (from the beginning of heating) is complicated [17]. An analysis of it shows that after some transition time t_1, the temperature in the inside of the sample can be given by

$$T = T_0 + bt - \frac{b(R^2 - r^2)}{4a} \qquad (4.8)$$

where R is the diameter of the sample cylinder (assumed to be the sample shape), r is the distance from the sample centre to the point where the thermocouple is located, and a is the temperature transfer coefficient of the sample.

Thus, after the transition time t_1, the sample temperature begins to rise at the same speed b as that of the heating source (this is called a quasi-stationary regime) [17]. The precision of this temperature rise is given by the linearity of the temperature programming device. Contemporary temperature programming devices can produce a linear temperature rise to better than 0.1%. Thus a linear sample-heating device removes several ambiguities of the pyrolysis temperature caused by different TRTs and different final temperatures in impulse PyGC.

The sample temperature normally differs from that set for the reactor by the programmer (Fig. 4.14). However, this difference can be neglected for most PyGC

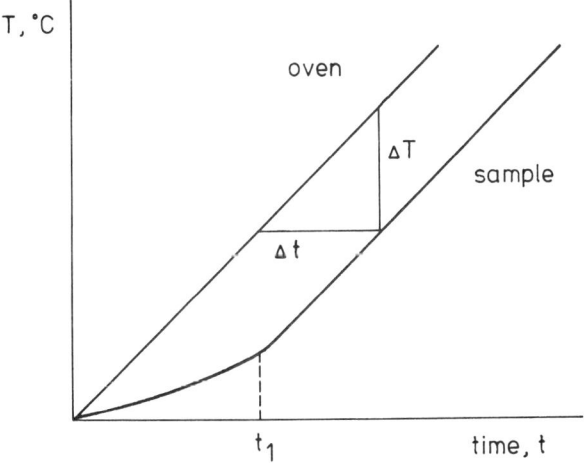

Fig. 4.14 — Theoretical curves for heater and sample temperature dependence on time.

measurements of polymer samples. For example, taking $R = 1$ mm, $r = 0$, $b = 10$ deg/min and $a = 0.004$ cm^2/sec [17], the temperature difference between the sample and reactor will be $bR^2/4a = 0.1°$.

As will be seen in Section 5.1.3, heating at a constant rate is one of the basic principles of thermochromatography.

4.3 INPUT SYSTEMS

The input system must supply the desired waveform to the inlet of the chromatographic column, i.e. it converts an electrical signal into a chemical one. Although a certain function may be advantageous from a theoretical point of view, the nature of the chromatographic apparatus and the inlet systems available enables us to realize easily only those input functions that are a combination of a step or impulse-like function. The generation of other functions is quite complicated. For example, to generate a sine concentration input function, a train of equi-interval pulses was passed through a short inlet column that smoothed the impulse fronts and backs [18]. This input function was certainly not a sine-wave but could be considered as a first approximation to one because the inlet column removed most of the harmonics from the square wave.

Automatic sampling devices can be quite sophisticated, depending on the constructor's ingenuity. Several input systems used to generate complex input functions have been described in the book by Foreman and Stockwell [19]. The most interesting is an overpressured injection system that can easily be controlled by switching solenoid valves. In this system, the sample flows from the vessel to the chromatographic column through the capillary tubing. The volume of the sample, V, that flows through the capillary column during the time interval Δt is given by the Hagen–Poiseuille law:

$$V = \frac{\pi d^4}{128\eta} \frac{\Delta p}{l} \Delta t$$

where Δp is the pressure drop in the capillary column, d is the column diameter, l is the length of the capillary and η is the sample viscosity. The overpressured device designed by Boer is shown in Fig. 4.15 [19]. If the solenoid valves V_1 and V_3 are closed and V_2 open, the sample is introduced into the column. After introduction of the sample, the valves are reversed in status, the overpressure is led to the atmosphere and the capillary C is continuously flushed with the hot carrier gas, thereby reducing to an insignificant amount any memory effect between the samples. Overpressure injection systems have several advantages: the program control can be quite flexible and the input impulse duration can be varied to a large extent to obtain a suitable amount of sample. These systems have been used in preparative chromatography and also headspace analysis. This sampler is good if the sampling frequency is not very high.

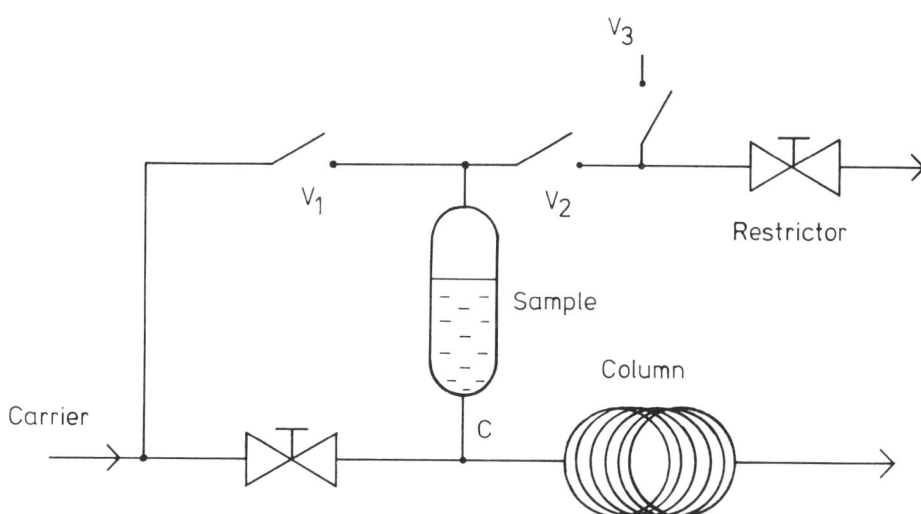

Fig. 4.15 — An overpressured sampling device [19]. V_1, V_2, V_3 — solenoid valves, C — capillary. (Reproduced with permission, from H. Boer, *J. Sci. Instrum.*, 1963, **40**, 121. Copyright 1963, Institute of Physics).

4.3.1 The effect of physical and chemical factors on the flow

Physical or chemical effects on the flow can cause interesting changes in it depending on their nature. This approach to sampling has been developed by Phillips *et al.* [1,3]. The results are promising. These types of sampling device have an important advantage over other input systems (mechanical and pneumatic valves) because they do not involve any mechanical part (not even solenoid valves) and can be built very compactly. Their interfacing to the computer is easy. As will be seen below, a particular feature of these devices is that they are problem-specific. This is because the physical or chemical modulation acts differently on different compounds in the flow, and not uniformly, as in the case of sampling by mechanical valves etc. The specificity of the modulation effect can be considered an advantage or disadvantage depending on the situation. For example, if only one target compound is of interest and it is possible to find a specific physical or chemical modulation mechanism that acts only on that component, it is possible to determine this compound in the presence of others. However, the specificity of the input system is usually a complicating factor in chromatographic analysis.

The easiest way to modulate the concentration flow is to direct the flow through a common chromatographic column and change the column temperature. The solubility of a flow component in the liquid phase depends on temperature (temperature-dependence of the capacity factor). The column temperature may be changed by electrical heating, with a voltage applied directly to the metal column [20,21], or more elegantly by painting part of the capillary with electrically conducting paint [1], Fig. 4.16. When the flow enters the column-modulator, the flow components are distributed between the mobile and stationary phases as they are in any column. The modulator soon reaches a quiescent condition (a steady state). If an electric current

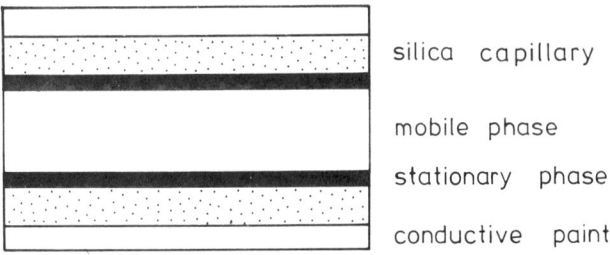

Fig. 4.16 — A thermal desorption modulator.

impulse is then passed through a thin film of the electrically conducting paint, it rapidly heats the column. A sudden temperature pulse releases any substance dissolved in the stationary phase, increasing its concentration in the mobile phase. The mobile phase carries the concentration pulse out of the modulator. In the second phase of the working cycle, the current is switched off, the modulator cools down, and now being 'empty', begins to absorb substances from the flow before reaching the quiescent condition again. The absorption process removes substances from the flow, thus generating a decrease in the concentration (a 'vacancy'). In this way, the modulator generates a concentration pulse in the flow, consisting of a positive and negative part, that resembles the derivative of a normal chromatogram.

The positive and the negative parts of the signal differ in duration. The velocity at which a given substance moves along the column is $u/(1 + k)$, where u is the mobile phase velocity and k is the capacity factor. When the modulator is hot the compounds released from the stationary phase move out of the modulator at the carrier speed. Thus, the duration of the positive part of the impulse is $t_+ = b/u$. Correspondingly, the generation of the negative part requires that t_- be equal to $b(k + 1)/u$. Here b is the modulator length. The modulator impulse is shown in Fig. 4.17 and the modulator chromatogram in Fig. 4.18. For the modulation impulse amplitude Phillips gives the formula [1] $A = C(k + 1) - C/(k + 1)$, where A is the amplitude of the modulation signal and C is the steady-state concentration in the modulator.

The modulator is used to introduce the random pulse sequence into the chromatograph. It follows from the modulator construction that a certain dead-time is necessary between pulses, to allow the modulator to cool down, i.e. during this dead-time the thermal pulse is not allowed. This dead-time depends on the thermal mass of the modulator, and for a modulator fabricated from a fused silica capillary the shortest impulse possible is 0.1 sec. This value is reasonable in working with capillary columns and does not result in too much instrumental broadening of the chromatogram peak. Several other characteristics of thermal desorption modulators and the theory have been reported [1].

The thermal desorption modulator has also been used in capillary liquid chromatography [22]. These columns accept substantially smaller quantities of sample than do packed LC columns, and the column sample-capacity limit may become less than the detection limit. Hence, it is sensible to apply the multiplex

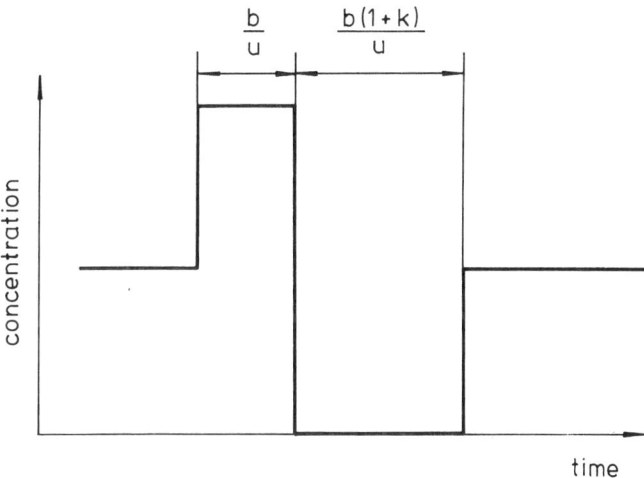

Fig. 4.17 — Impulse response-form of thermal desorption modulator.

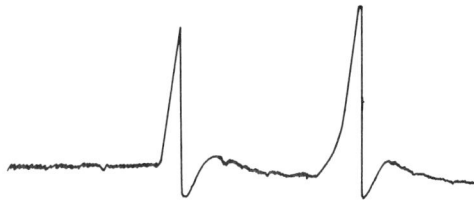

Fig. 4.18 — Characteristic chromatogram from a thermal degradation.

chromatographic technique to improve the capillary LC performance. Typically a modulator is made of the same material as the packed capillary LC column, e.g. 0.35 mm i.d. fused silica capillary packed with 7 μm Chemicosorb ODS/H placed before the packed column. This set-up allows us to minimize the injection port substantially in comparison with common mechanical valves. However, care must be taken not to heat the modulator above the solvent boiling point.

Thermal decomposition is a simple technique that lends itself well to selective modulation. It can make use of the same arrangement as thermal desorption modulation. The application of the pulse to the thermal decomposition modulator degrades thermally labile compounds in the sample stream, thus removing the samples of interest from the stream. The resulting chromatogram contains negative peaks. This idea was applied to an artificial sample stream containing several saturated and unsaturated hydrocarbons [23]. The latter are thermally degraded in the modulator and the corresponding peaks will be seen on the chromatogram. A typical chromatogram obtained in this way is shown in Fig. 4.19. The negative peaks followed by positive ones in the chromatogram are within the retention time interval 0.22–0.5 min. These peaks are assigned to the carbon monoxide 'vacancy' and carbon dioxide (a positive peak). When the modulator is hot, the carbon monoxide impurity is oxidized to carbon dioxide by the trace amount of oxygen in the stream. Thus, carbon monoxide is removed and carbon dioxide added to the stream, resulting in a positive peak with a 'vacancy' immediately preceding it.

Other chemical reactions can also be successfully used in selective modulation. The amount of methane in the ambient air has been modulated by a modulator containing silver(I) oxide [3]. The modulator selectively catalyses decomposition of the methane. Photochemical modulation of a sample flow has also been used [24]. Visible light was the modulation source and a dye, alizarin, was used as the stationary phase in the modulator. The light caused a photochemical reaction between the stationary phase and an aniline derivative used as the sample, and the modulation of the light intensity enabled modulation of the sample flow.

The use of an electrochemical modulator in LC has also been investigated [25]. A precolumn electrochemical cell was used to generate chemical concentration signals by random switching between two positive cell potentials. In this way, the chemical state of an analyte may be modulated between its standard and oxidized forms. The signal generated by the electrochemical modulation consists of two parts. One is a vacancy resulting from removing the analyte in its standard form. The second is one peak (or more) resulting from the oxidation product(s) of the analyte. Aniline was used as test substance. In this paper, a remarkable improvement in detection limit of three orders of magnitude was obtained by using the multiplex sampling technique. This improvement, however, should partially be assigned to factors other than the multiplex advantage (e.g. to the higher sensitivity of the detector to the oxidized product form).

Nowadays, the quality of chromatograms obtained by modulators is much better than that of those published earlier. Modulators have great possibilities where instrument compactness, strength, simplicity and lightness are of prime importance, e.g. in space studies. Their application when physical separation is necessary (e.g. two-dimensional separations) is limited, however. The most widely used physical factor for modulation, temperature, is, however, quite a complicated parameter to control. On the other hand, the equipment for temperature modulation is very easy to build. Further means of sample concentration modulation should be searched for and investigated. Up to now, the use and investigation of modulators are at an academic level and no commercial products are yet available.

4.3.2 Pneumatic flow switches

The first pneumatic flow switch, invented by Deans, has already been described in Section 3.3.2 in terms of its advantages and disadvantages. The pneumatic flow switch can be regarded as a special kind of modulator where the modulation effect is achieved by changing the flow pressure at several points in the flow path, thus changing the direction of the flow through the corresponding elements of the switch. The basic idea of pneumatic flow switching can be explained according to the simple scheme shown in Fig. 4.20. A similar arrangement has been described in [26]. The sample flow is directed through the restrictor, R_2, by applying control pressure to the

Sec. 4.3] **Input systems** 123

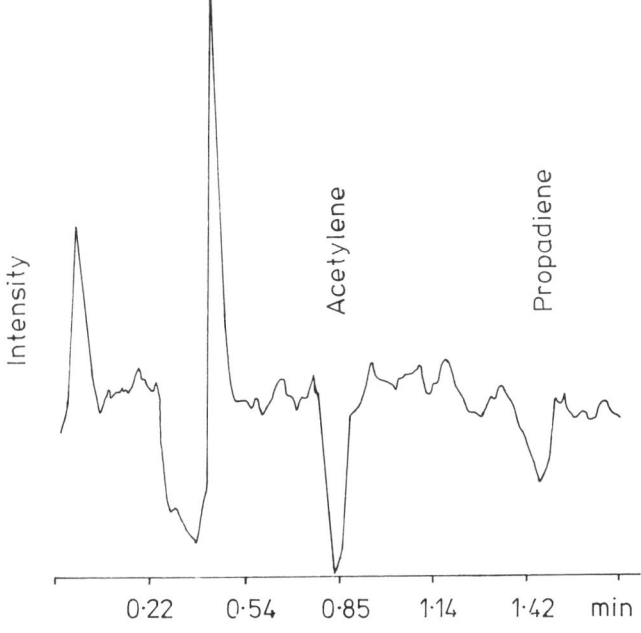

Fig. 4.19 — Characteristic chromatogram from a thermal degradation modulator. (Reprinted with permission from J. R. Valentin, G. C. Carle and J. B. Phillips, *HRC&CC*, 1983, **6**, 621. Copyright, 1979, Hüthig Verlag).

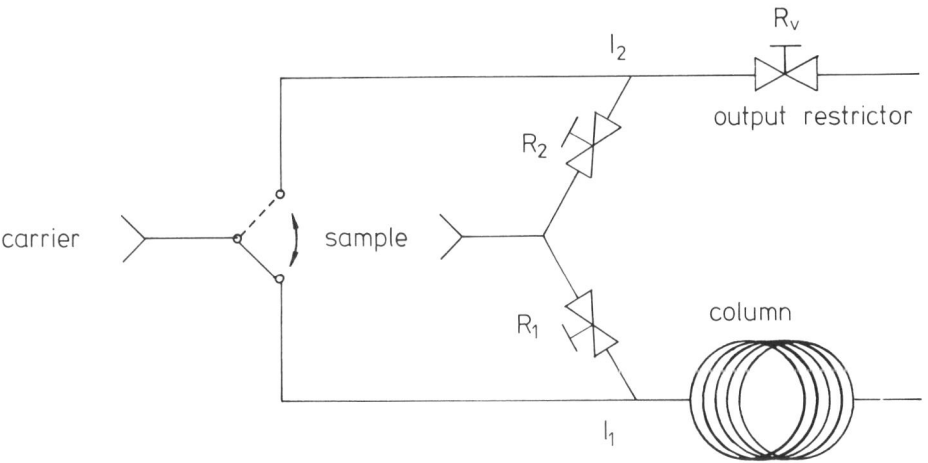

Fig. 4.20 — Pneumatic flow switch [26]. R_1, R_2, R_v, restrictors; I_1 and I_2 carrier inputs.

point I_1. The pure carrier flows through the column. If control pressure is applied to I_2, the sample flows through R_1 and the column. However, the switch works properly only if the carrier and sample pressures (and flow-rates) have been set up correctly at the corresponding points. Several modifications of this basic scheme can be found. The topological arrangement of Deans's first pneumatic flow switch is the same as that described in Fig. 4.20, despite the fact that Fig. 3.8 looks quite different. Below we shall consider some pneumatic switches described in the literature.

Another invention of Deans can be described as follows (Fig. 4.21) [27]. The

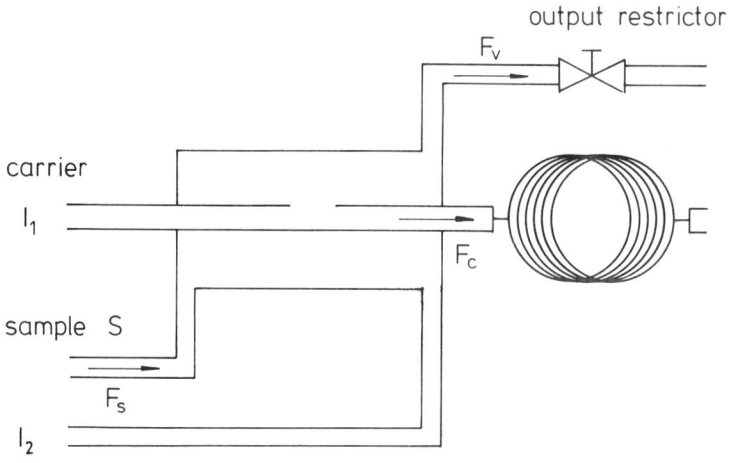

Fig. 4.21 — Pneumatic flow switch [27]. I_1, I_2 — inputs; F_v, F_s, F_c — output, sample and column flows.

switch is fabricated from two tubes, one inside the other. The smaller tube has a hole drilled in its wall. When pressure is applied to the input I_1, and I_2 is closed, part of the carrier gas flows through the hole and is vented. The sample does not enter the column. If pressure is applied to input I_2 and I_1 is closed, the sample enters the column through the hole. In this manner, sampling is achieved. To work properly, the rate of the volume flow through the switch must satisfy the condition [27]:

$$F_v > F_s > F_c \qquad (4.9)$$

where F_v, F_s and F_c are the flow-rates through the vent restrictor, the sample source, and the column inlet, respectively. The condition (4.9) severely restricts the available flow-rates through the system and is frequently considered to be a disadvantage of this device. Also, establishment of these flows is not always easy, although it seems to be trivial. Any changes in the chromatograph that affect system flow-rates require re-establishment of the latter. This is relatively easy when a thermal conductivity detector is used and the flow can be measured on fixed flowmeters, but if a flame ionization detector is used, changing the flow-rate can require an inconvenient demounting of the column.

Two advantages of these types of switch are especially remarkable: they can work at any temperature encountered in GC, and can be made within a few hours by the user himself. The latter feature is of prime importance, owing to the high cost of other switches and valves manufactured by instrument companies, especially if laboratory resources are limited.

There are no studies of this switch other than Deans's original work [27]. The reproducibility of the repetitive sample amount was 0.12% for work with an isothermal system and 0.5% when the switch temperature was programmed. We have been using this type of switch for several years, and find a reproducibility of about 0.3%. The switch performance was controlled by correlation chromatography [28], which is well-suited for testing input system reproducibility because, as shown in Section 2.2.4, it is very sensitive to variation in sample amount, and any change in amount of sample results in correlation noise. In [28], the sample was a stream of ethanol from a diffusion cell. Pseudo-random sampling with a Deans switch gave a chromatogram with the expected noise level, which means that no additional noise was generated by the sampling system.

No studies of the influence of the switch geometry on switch performance are available. The original switch [27] was constructed from 1/8- and 1/16-in. stainless-steel tubes and Swagelok couplings. We have used a similar construction made from tubes [28]. The other important parameter in characterizing the device is the switching speed, which determines the amount of sample that can be introduced by the switch, and also the applicability of the switch to high-speed separations. Deans reports that about 0.1-sec pulses can be successfully introduced into the chromatograph. That value is limited by the solenoid valve actuating speed, but evidently also depends on the switch geometry. This problem needs further study.

Valveless flow switches are usually pressure-controlled. This may become a problem when temperature programming of an analytical column is necessary [29]. Using constant-flow controllers in the flow-path to set up proper flow-rates permits the utilization of wider temperature range programs, but unlike pressure-controlled systems, there is some baseline disturbance because constant-flow controllers need a few seconds to stabilize when valves are switched on.

In [30] capillary columns in a two-dimensional system were connected through a specially designed manifold acting as a pneumatic flow switch (Fig. 4.22). The manifold was fabricated from a piece of nickel tubing (1/16 in. o.d., 0.50 mm i.d., 150 mm long). A narrow indentation was made in the nickel tubing, to serve as a flow restrictor. If $P_e < P_f$, the sample flows to the vent (through the monitor FID) and does not enter the analytical column. If $P_e > P_f$, the sample flow is directed into the analytical column. Several characteristics of this switch were measured [30]. The band broadening caused by introducing this manifold into the capillary column system was found to be negligible. The switching speed from the 'backflush' to the sampling mode was found to be 0.3 sec, and 0.8 sec for switching in the opposite direction. It was observed that an increase in the flow resistance due to the switch indentation resulted in a slower response of the system.

The switches described in [27–30] were home-made. The fabrication of the pneumatic switch requires some skill, enthusiasm and a good mechanical workshop,

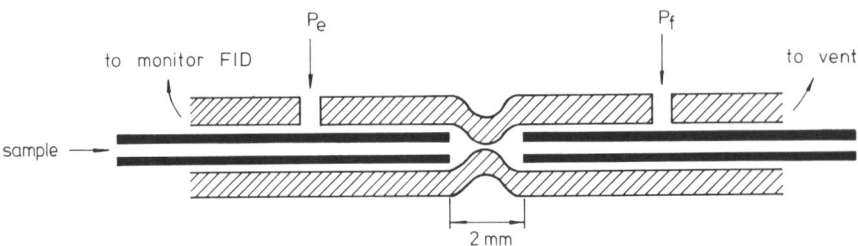

Fig. 4.22 — Pneumatic flow switch. P_e, P_f — pressures. (Reprinted with permission from H. Brotell, G. Rietz, S. Sandqvist, M. Berg and H. Ersson, *HRC&CC*, 1982, **5**, 596. Copyright 1982, Hüthig Verlag).

and unfortunately is not always possible for the chromatographer. The ideas described in [27–30] have been further developed in several pneumatic switches manufactured by some instrument companies. The hardware has been specially designed for two-dimensional separations and constructed in such a way that it is possible to use the units with almost any existing chromatograph. This should permit wider use of two-dimensional chromatography. Below we describe the three-column switching hardware now commercially available.

The Scientific Glass Engineering Co. (SGE) multidimensional conversion system for GC [31] is based on the pressure-balancing technique first proposed by Deans. It enables the following operations to be automatically controlled: heart-cutting, back flushing, fore-flushing, and cold-trapping. All but three components of the SGE multidimensional conversion system are positioned in a free-standing pneumatic control module (solenoid valves, flow restrictors and meters, pressure regulators and gauges). The only modification required to be made to a chromatograph is the mounting of the midpoint restrictor (at the juncture between the precolumn and analytical column), a cold-trap T-junction, and a pneumatic shut-off valve. The pneumatic control module is connected to the corresponding points in the GC oven by 1/16 in. transfer lines. The midpoint restrictor is fabricated from glass-lined stainless-steel tubing in order to maintain an all-glass column-to-column pathway. Low dead-volume of the midpoint restrictor is achieved by using a low dead-value butt connection with a graphitized two-hole ferrule on the precolumn side, and a swept tube connection to the second column on the opposite side. The switch utilizes a pin-point restrictor on the wall of the glass to provide the required restriction between the columns. The SGE multidimensional conversion system gives a relative standard deviation of 0.03% for peak retention times and 1.5% for peak area (these are mean values for several modes of work).

The Siemens SiChromat 2 chromatograph is particularly suited for use with the valveless column-switching technique on the 'live' principle. The term 'live', proposed by the company workers, denotes that switching is realized by reversing the pressure drop on the restrictor [32]. The switch is shown in Fig. 4.23. The system works with slight pressure differences between both ends of the T-piece, generated by means of two gas lines (A, B). The gas flows through these lines are adjusted by using the two needle valves, NV2 and NV3. Any gas flow from one end of the switch

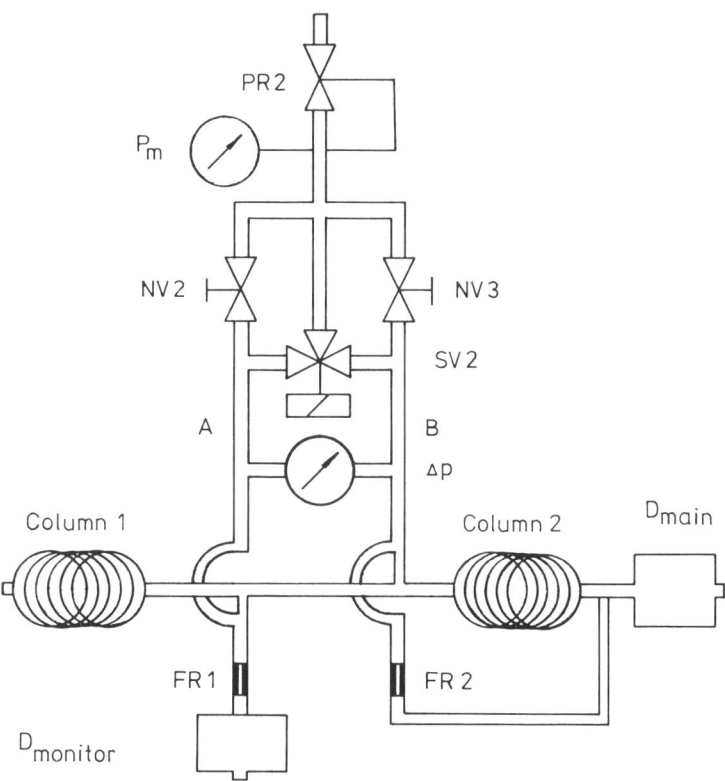

Fig. 4.23 — 'Live' column switching unit. (Adapted by permission of Siemens Aktiengessellschaft).

to the other must pass through a platinum–iridium capillary mounted in the centre and inserted loosely into the two capillary columns (Fig. 4.24). The narrowest inner diameter of the capillary column that can be used is 0.3 mm. The direction of the flow inside the platinum–iridium capillary connecting the precolumn and the analytical column can be changed by opening and closing the external solenoid valve, SV2. When a higher pressure is used in line B than in line A, the effluent from the precolumn is directed to the vent. The sample is directed to the second column by reversing the pressure difference across the switch. Thus, the idea of 'live' switching is the same as that already described in Fig. 4.22. Although many chromatograms produced by the SiChromat 2 have been published, no data about switching speed and peak area reproducibility are available in the literature.

The design of the Chrompac Co. 'MUltiple Switching Intelligent Controller' (MUSIC) [33] is quite similar to that already described. The MUSIC is a modular system that can be built into every modern gas chromatograph. The system consists of four components: a microprocessor-based programmer unit, a pneumatic unit (a solenoid and needle valve, pressure and flow controller) controlled by the program, a

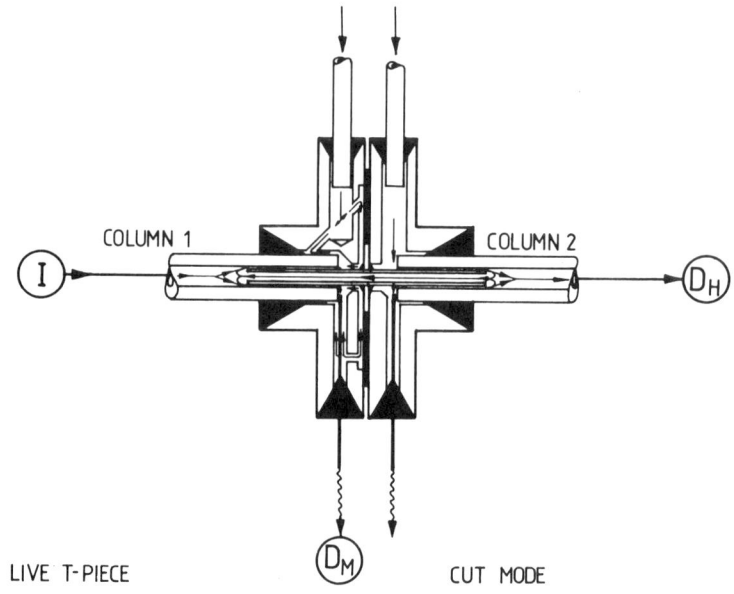

Fig. 4.24 — 'Live' T-piece. (Reprinted with permission of Siemens Aktiengesellschaft).

column module (for column connecting and flow switching), and a cold-trap. The functions of all these parts have already been described. No data are available about the performance of the system. The basic scheme of the MUSIC is shown in Fig. 4.25.

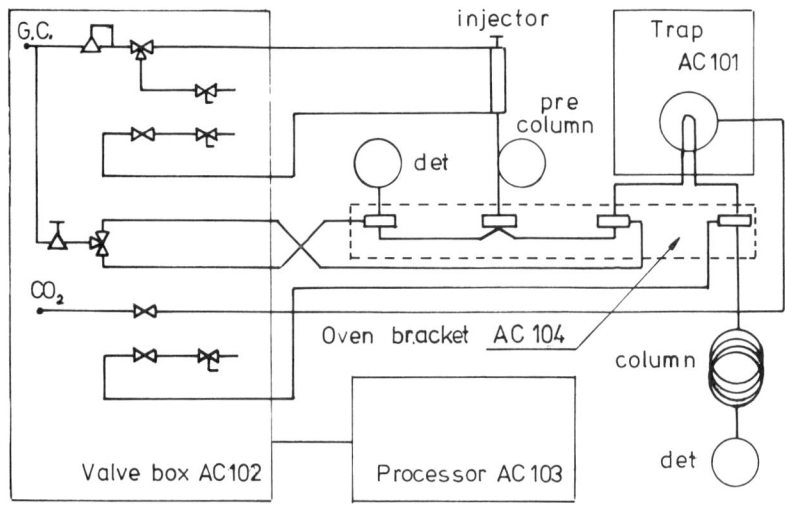

Fig. 4.25 — Flow scheme for the MUSIC. (Reprinted with permission of Chrompack International B.V.).

4.3.3 Flow switches based on fluidic elements

Fluidics is a branch of computer science and automatic control where the gas (or liquid) flow is considered to be a signal carrier [34]. Equivalents for all known electronic elements (e.g. triggers, amplifiers, shift registers, etc.) have been designed for fluidic systems. As the signal carrier in the fluidic systems and chromatography is the same, the idea of utilizing some fluid elements in chromatography is quite natural. However, the only fluidic element used in GC so far is the OR/NOR gate shown in Fig. 4.26.

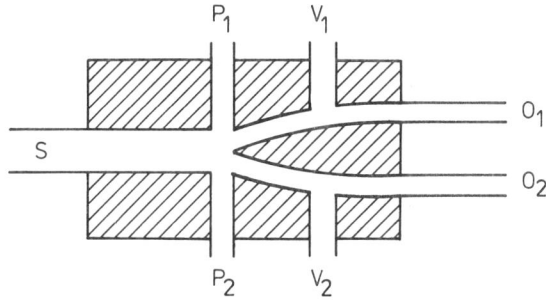

Fig. 4.26 — Fluidic logic gate: S — sample input; P_1, P_2, controlling pressures; O_1, O_2, outputs; V_1, V_2 vents.

The device is a ceramic or plastic piece grooved according to the geometry shown in Fig. 4.26. S is the sample input, P_1 and P_2 are inputs where a control pressure is alternately applied, O_1 and O_2 are outputs to the column and the vent. Two additional vents V_1 and V_2 are also necessary for a proper functioning of the gate.

Similarity between the devices shown in Figs. 4.20 and 4.26 is evident. The fluidic switch seems to be the ultimate in the development of the pneumatic switch, where all elements have been fabricated in one piece. However, the analogy is illusory and reflects a general structure of the chromatographic input system. The main difference between the pneumatic and fluidic flow switch is that the former works under laminar flow conditions whereas the fluidic gate requires turbulent flow. The vents V_1 and V_2 are necessary for generating these conditions. When the sample enters through the input, it is attached to the wall of the channel O_1, V_1, if pressure is applied to the control input P_2. This is due to the Coanda effect [34]. If pressure is applied to the control input P_1 the sample flow switches to the channel O_2, V_2. This can be performed very rapidly. Switching the flow from one position to the other and back can be performed within 10 msec [35]. Thus it is possible to generate very short initial chromatographic zone band widths, and the use of fluidic flow switches is very attractive in high-speed gas chromatography, as already seen in Section 3.2.2. The signal carrier in conventional fluidic devices is air fed to the system by an ordinary compressor. In chromatography, the carrier is usually purified helium, hydrogen or nitrogen. The first is too expensive to be consumed in large amounts, the second is a hazardous explosive material, and the last is not very good for use with certain detectors (e.g. TCD).

Three problems are evident when using these devices: (1) the switch consumes a large amount of the carrier/sample gas; (2) it is difficult to interface the fluidic gates, which have a low impedence and need a high flow-rate, to the chromatograph columns, which have a high impedence and require a low flow-rate (the gate requires a flow of 2–5 l./min); (3) the amount of sample injected by the switch approaches the detection limit of most modern chromatographic detectors. Further, the fluidic switch adsorbs some of the sample on its walls and, generally, it can be used only at room temperature. These drawbacks have restricted the use of fluidic devices. Below we shall consider some ways of overcoming these difficulties.

The solution of the first two problems has been suggested by Gaspar *et al.* [35, 36]. They demonstrated that a fluidic switch could be operated at the high pressure required at the column input by enclosing the gate in a vessel elevated to column pressure. The Reynolds number in the central channel, which should be several thousand for turbulent flow, is a function of the mass flow-rate irrespective of pressure [35]. Also, when several on/off switching valves are included in the gas supply lines of the gate, it is possible to fire the gate only at the time of injection, during a 10-sec period. A schematic diagram of the gas circuit is shown in Fig. 4.27.

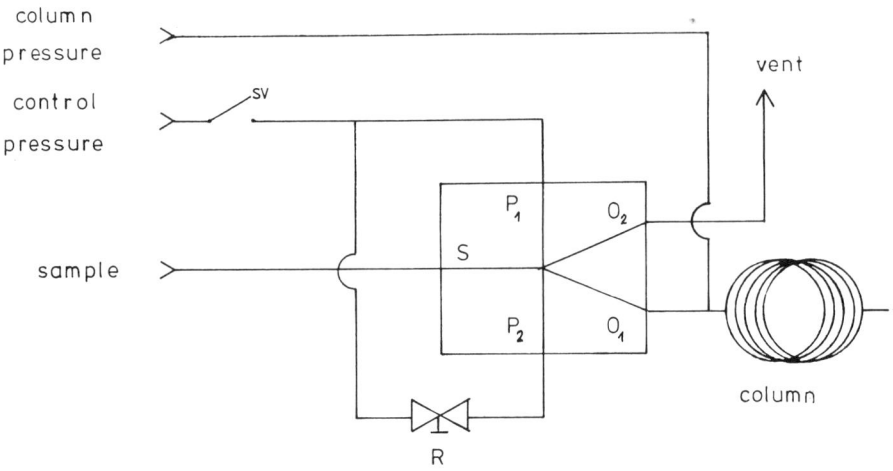

Fig. 4.27 — Fluidic sample switch for high-speed chromatography (adapted from [35]).

The fluidic input system operates as follows. Once a steady state has been achieved after opening the on/off valves, the solenoid valve SV is triggered. A pressure pulse appears at P_1, and the same pressure pulse travels along the delay line through the restrictor R and appears at P_2 after the delay time Δt that determines the injection duration. By adjustment of the restrictor, the injection time Δt can be regulated. A similar system has also been used by Schutjes *et al.* [37]. In this work, the carrier gas through the fluidic gate was nitrogen with a flow-rate of 1.5 l./min and the sample was injected manually into a common chromatographic injector included in the sample channel S. The resulting vapour band, having a duration of several seconds, was carried to the fluidic gate by the nitrogen flow.

The amount of sample injected by the fluidic switch is very small (in the range of a few pg [37]) and the detector noise becomes a problem in peak detection. Applying correlation chromatography (CC) to improve the signal to noise ratio was promising at the beginning [38], but difficulties arose in the application of CC to high-speed chromatography with fluidic switches [39]. Because of its extreme sensitivity to system deviations from the required ideal performance, CC is a good indicator for estimation of input system quality. The required input system quality was not achieved by Annino and Leone [39]. They proposed an ensemble-averaging technique to improve the signal to noise ratio in high-speed chromatography. Ensemble-averaging, however, is a signal collecting method, and to improve S/N by a factor of $\sqrt{n}$, n chromatograms must be run sequentially. If, for example, a tenfold increase in S/N is required, a hundred chromatograms should be recorded and averaged. Unfortunately, this is no longer high-speed chromatography. An important advantage of the CC method over ensemble-averaging is saving in time. Hence, if the importance of high-speed chromatography increases, it would be prudent to improve the construction of fluidic input devices to meet the requirements of correlation chromatography.

The fluidic switches are being thoroughly studied by a small group of investigators [35–39] and they seem to be the only possible input devices for high-speed chromatography today. Several problems related to the application of these devices remain and need further study, however, and no commercial systems are as yet available.

4.3.4 Mechanical and solenoid valves

Mechanical valves have long been in use and their construction and properties have been well documented, so we shall give only a brief description of them. The valve usually includes a switching unit which has properly grooved flow pathways, an actuator which operates the switching unit by compressed air, and solenoid valves to control the air flow to the actuator. In the manual switching mode, the actuator and solenoid are replaced by a shaft. Popular in the early years of chromatography, plunger-type valves have been displaced by rotor-type ones.

The flow path of a common six-port valve is shown in Fig. 4.28. The major problems met in the beginning with mechanical valves were lack of inertness, leak-tightness, and mechanical stability, but these have now been overcome by valve manufacturers. According to the specifications of the Valco Instruments Co., a valve can inject samples in the volume range of 0.2–5.0 ml. Inertness is achieved by using various metals in the valve bodies (nickel/chrome, Hastalloy C, tantalum). With compression-moulded special polymers (PTFE and polyimide resin) as rotor material and matching of the thermal expansion coefficients of the valve body materials, and valves can work in the temperature range from -198 to $+300°C$ with minimum leakage. The valves can perform a great many switching cycles without failure. Replacing the seal assembly restores the valve to almost new condition. In liquid chromatography, valves do not require temperature stability, but may have to be

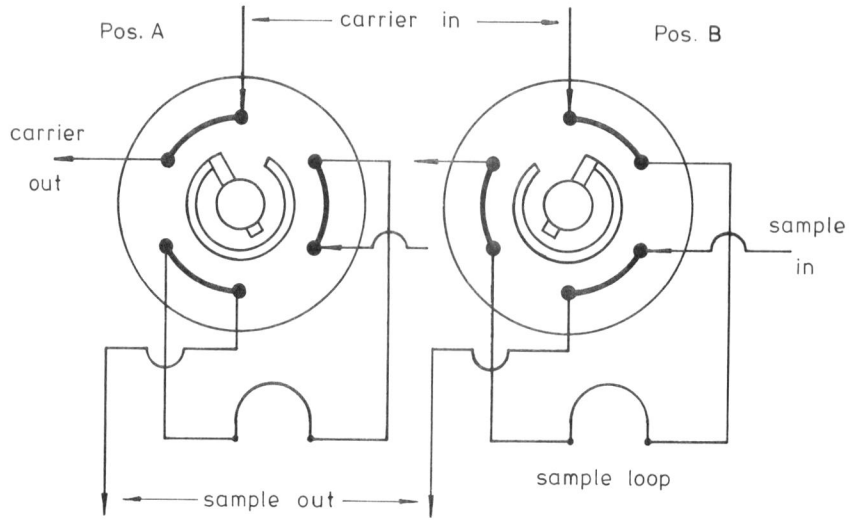

Fig. 4.28 — A six-port two-position mechanical valve.

made to withstand high pressure (up to 35 MPa). Multiposition valves have been developed, and may have 4–16 sequential positions instead of only 2. The construction and applications are well described in the manufacturers' literature [40].

Mechanical and pneumatic switching valves have already been compared in Section 3.3.2. The possibility of straightforward construction of a gas flow-path in complicated switching systems makes mechanical valves attractive, whereas pneumatic flow switches tend to behave in an unpredictable and mysterious way. The air actuator of the mechanical valve permits the application of a controlled constant torque to the rotary parts. It contains a pneumatic piston that transfers the air pressure in the actuator to the rod turning the valve rotor, by a transmission mechanism.

Solenoid valves are necessary components in contemporary chromatographic systems (an exception may be flow modulators). They transduce electrical signals from the controlling computer into pneumatic signals. The solenoid valve (Fig. 4.29) consists of a bar made of magnetic material, and a solenoid. The bar is kept in one position by a spring. If electric current is applied to the solenoid coil, it moves the bar to the other position. Thus, the solenoid valve is a two-positional unit that can change the gas flow from one output to the other by moving the bar. To drive the air actuator of the mechanical valve, two solenoid valves are needed.

The whole system (a valve, an actuator, and a solenoid) has a switching speed of about 0.2 sec. However, by using large-diameter connection tubes (1/4 in.) with helium as actuator gas, it is possible to lower this to 7 msec (see Section 3.3.2). The solenoid valves themselves have a switching time of about a few msec.

Sec. 4.3] Input systems 133

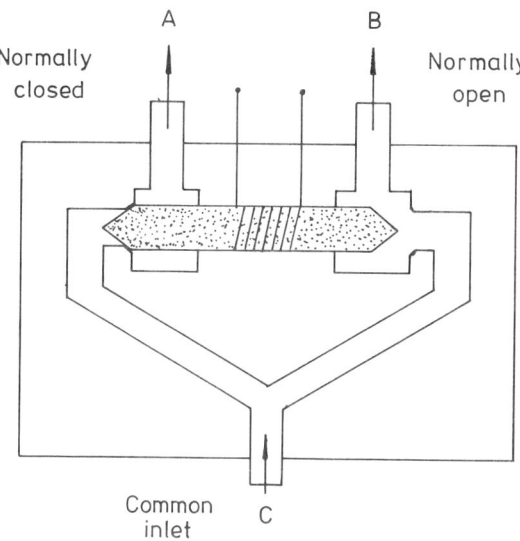

Fig. 4.29 — Basic scheme of a solenoid valve.

4.3.5 Autosamplers and robot sampling

Autosamplers have been designed for the automatic introduction of liquid samples in GC or HPLC. They include a round table with vials along its circumference. The table turns step by step, with transfer of the contents of a particular vial by a complicated mechanical injection device. When an autosampler is used, a relatively long sequence of events is involved: (1) introduction of the sample and solvent into a vial; (2) capping of the vial with a septum at the capping station; (3) placing the capped sample vial in the autosampler; (4) activating the autosampler to effect sample injection. Such autosamplers are not useful in GC applications where time is an important variable, e.g. in kinetic studies.

Laboratory robots, the latest invention in analytical chemistry, are becoming increasingly popular as sample preparation devices. For liquid sample introduction in GC and HPLC they are an attractive alternative to autosamplers. A robot can be programmed to perform a sequence of actions of a complexity that is limited only by the ingenuity of the user.

A successful adaption of the Hamilton GC syringe to the Zymate robot syringe hand was reported by a group of investigators at Duke University [41]. The syringe was connected to a special syringe hand of the robot. The positioning accuracy of the Zymate robot is approximately 2.5 mm. Most GC injection ports have an internal diameter of 1.0 mm or less. To make the injection port robot-friendly, the opening of the injection port was enlarged to 2 mm. A syringe guide was then fabricated from a polyethylene disposable pipette tip and a temperature-resistant silicone cement. The technical modifications described above are simple and straightforward and can be made by any chromatographer. Much more precise syringe adapters and needle guides could be fabricated to achieve the same result. The injection precision by this system was 1.5%, and the error was mainly due to the Hamilton syringe itself. The robot injection system was used for a series of trans-esterification studies.

The flexibility of the robot system is due to the microprocessor controlling the robot. The robot can be programmed to perform a sophisticated sequence of actions that would be impossible by a human. Multiple injections required by boxcar type experiments (see Section 3.2.4) for studying fast kinetics could easily be performed by a robot injection system. With big computers and special programming languages (e.g. LISP), expert systems can be produced that teach themselves and develop new analytical methods [42]. There are good prospects for these systems in chromatography.

A drawback of the robot systems is the relatively high price compared to that of autosamplers or other sampling systems. However, for routine application (e.g. in an industrial laboratory) the robot could pay for itself within a few months, and for non-routine applications, (e.g. in performing new types of experiment realizable only by using robots) current trends strongly promote the use of robots.

4.3.6 Summary of input systems

To give an overview of the input systems described in Section 4.3 their characteristics are summarized in Table 4.4.

4.4 CHROMATOGRAPHS: DEMANDS OF MULTIPLE INPUT

A modern chromatograph is a sophisticated device usually controlled by a microprocessor, and all the necessary parameters (carrier speed, temperatures, etc.) are set up by the chromatographer from a keyboard. Also, several other parameters are introduced through the keyboard, e.g. sample component names and response factors (if known), temperature or solvent programs. Time–event sequence programming is possible, i.e., at several predefined times the computer commands the chromatograph to perform some action (e.g. change some of the parameters).

Gas chromatographic separation is done mainly in fused silica capillary columns, the performance of which has greatly improved during the last decade, but some analytical problems still require the use of packed columns, e.g. inorganic gas separation. For sample introduction, a number of ingenious injection devices have been developed. They are, however, useful only for manual sample introduction and are not of interest for multiple-input computerized chromatography.

Only a few detectors are in wide use, although the number of detectors developed for gas chromatography is large. The two most popular detectors are the flame ionization and thermal conductivity detectors. The most popular element-selective detectors are flame photometric, electron capture, electrolytic conductivity, and thermionic detectors. Mass and infrared spectrometers are in use as GC detectors.

High-performance liquid chromatography has now approached the level of sophistication achieved in all elements of GC except the detectors. Good pumps and a wide variety of high-resolution columns are now available. However, since the carrier is a liquid, HPLC detectors have limited sensitivity, selectivity and versatility. The most popular detectors are refractive index, ultraviolet, and fluorescence detectors.

Table 4.4 — Input system characteristics

No.		Sample type	Reproducibility (%)	Switching speed* (sec)	Cost†	Advantages	Disadvantages
1	Flow modulator	gas, liquid	no data	1–5 (0.1)	negligible	specificity; no moving parts	specificity
2	Pneumatic flow switches						
	(a) Deans type	gas	0.1–0.3	0.1	negligible	no moving parts; no temperature limits	requires careful design and experience of use
	(b) Commercial (MUSIC, SiChromat, SGE)	gas	2	1	medium		
3	Fluidic gate	gas	no data	0.010	low	fast switching speed	large sample consumption, requires skill to work
4	Mechanical valves	gas, liquid	no data	0.2 (0.007)	medium	straightforward construction of flow-path	leaks, temperature limits
5	Robot sampling	gas, liquid	1.5	1	high	flexibility	high cost

*The numbers are mean values, the numbers in parentheses are extreme values obtained by special equipment.
†Cost is relative to that of a traditional chromatograph.

Data-handling systems of modern chromatographs are able to print out sample composition in the form of a table of peak retention times and relative areas. Chromatograms can be stored in a file or on a disc, and processed later.

A chromatograph-controlling and data-handling computer can be directly used in several applications of multiple-input chromatography. Because of the possibility of real-time sequence programming the computer is able to send out electrical signals that can control external relays e.g. valve switching timers. Also, the controlling microprocessor can accept signals from external devices. In this way, a modern chromatograph can successfully control (and be controlled) by a laboratory robot without modifications to either of them.

However, the 'intelligence' of a built-in microprocessor is quite limited, and for a more sophisticated and special use, an external computer system is necessary.

Depending on the form of the multiple input, different modifications of the chromatograph are necessary. Two-dimensional separation has no special requirements with regard to chromatographs (except that there should be two of them). If it is possible to use only one chromatograph in LC then the monitor detector must be able to withstand pressure in the system.

It is possible to use any chromatographs in correlation chromatography, but the injection port should be removed and replaced by the multiple sampling system. CC requires a detector with a high noise level. The output voltage of FID or a TCD is of the order of a few μV, as is the analogue-to-digital converter resolution (the least significant bit value). Measuring at the noise level thus leads to a large quantization error which, although suppressed by a CC technique, causes unnecessary increase in noise level in the decorrelated chromatograms. A possible solution is to amplify the detector output with an instrumental amplifier to match the available ADC resolution better [1, 5]. The problem of low-level output voltage is especially important when older types of TCD are used. Modern TCDs are equipped with an output voltage amplifier and designed to meet the needs of computerized data handling.

High-speed chromatography requires the very best of chromatographic equipment. As already mentioned in Section 3.2.2, the main problem in high-speed chromatography is the instrumental contribution to band broadening. It spoils the expected resolution. The sources of the instrumental contribution to band broadening are the injection device, connections, detector volumes and the detector time constant, the last being the most important factor in high-speed chromatography. FID, the only detector still used in this field, has a response time of 50–60 msec. High-speed gas chromatography demands a much lower value of the time constant (10 msec) and thus requires a special amplifier or the rebuilding of a standard FID amplifier to lower its time constant. The capillary columns used in high-speed chromatography have a very narrow inner diameter (30 μm). They are not commercially available and should be specially ordered or made by the chromatographer.

4.5 COMPUTERS: DEMANDS OF MULTIPLE-INJECTION CHROMATOGRAPHY

It is difficult to write about computers: every statement will be out of date by the time it is read. Large minicomputers that two decades ago were attached with great efforts to scientific instruments are now transformed into one-chip microprocessors included in the instruments [43]. In modern analytical instruments the distinction between the computer and the instrument disappears because the instrument can no longer function as an independent device. Ultimately this has led to the development of new types of analytical methods and instruments based on principles that would not have been practical or possible without the involvement of computers. Many such measurements have already been described in this book. In the modern analytical instrument the keypad control panel provides simple function selection by being directly labelled with the commands and messages. Pushing the corresponding buttons instructs the dedicated computer to perform a certain action. How this separation from direct involvement with the instrument could frustrate a knowledgeable operator has already been mentioned in the Foreword.

Although these 'smart' controllers add an extra convenience for routine analysis they give little help in the development of new computerized analytical methods. Method development requires sophisticated programming, and acquisition, storing and interpretation of data, and thus does not require a dedicated computer. However, if the development stage is over, all the programs can be burned into a ROM and a new instrument with smart processor appears.

For computerized multiple-input chromatography the contemporary personal computers seem to be best. These computers frequently have slots for interfacing boards that allow them to solve different real-life problems. This partly explains the tremendous success of the Apple II computer. Many companies are making interface cards for the Apple II computer to enhance its possibilities and power. Apple II also has input/output relays that can be programmed by software. These relays are ideally suited for the switching of sampling valves.

For application in a scientific instrument a computer must be able to acquire data from detectors and sensors and control instrument operations (at least some of them). The circuits that provide these actions are called interfaces. The ease of interfacing determines the degree of involvement of the computer in the analytical instrument and thus strongly affects the sophistication of the computerized experiment. In Fig. 4.1 two basic interfaces are shown: the analogue-to-digital converter (ADC) and the digital-to-analogue converter (DAC). They convert a continuous electrical signal into a digital word, that will be sent to the central processing unit of the computer, and vice versa. Thus, all the information processing is done by electrical signals. Although this is not the only possibility and totally pneumatic computers (and chromatographs) are known, the electrical circuit remains the most common information carrier today.

In early applications of computers the interface cards were custom made, highly specialized, complex, and expensive. Today single-chip ADCs and DACs are commercially available. Chromatography, owing to its signal specificity, presents special requirements to the ADC. For a flame ionization detector the dynamic range is from 10 μV to 10 V, so if the least significant bit value is taken as equal to 10 μV a 24-bit ADC is needed to cover the dynamic range. The sophistication and price of this device is comparable to that of the personal computer. Another approach is to use an ADC that outputs a floating point number but this is also a very specialized and expensive device.

REFERENCES

[1] J. B. Phillips, D. Luu, J. Pawliszyn and G. C. Carle, *Anal. Chem.*, 1985, **57**, 2779.
[2] Tj. Th. Lub, H. C. Smit and H. Poppe, *J. Chromatog.*, 1978, **149**, 721.
[3] J. R. Valentin, G. C. Carle and J. B. Phillips, *Anal. Chem.*, 1985, **57**, 1035.
[4] J. E. Lovelock, *Anal. Chem.*, 1962, **33**, 162.
[5] M. Koel, M. Kaljurand and E. Küllik, *Proc. Estonian SSR Acad.*, *Chemistry*, 1983, **32**, 125.
[6] A. P. Altshuller and I. R. Cohen, *Anal. Chem.*, 1960, **32**, 602.
[7] D. K. Kollerov, *Methodical Background of Gas-Analytical Measurements*, pp. 199–226. GOST, Moscow (in Russian).
[8] F. P. Scaringelli, A. E. O'Keefe, E. Rosenberg and J. B. Bell, *Anal. Chem.*, 1970, **42**, 871.
[9] P. Steingaszner, *Microreactional Gas Chromatographic Methods*, in L. S. Ettre and W. H. McFadden (eds.), *Ancillary Techniques of Gas Chromatography*, p. 13, Wiley–Interscience, New York, 1969.
[10] J. Coca and H. Langer, *ChemTech*, 1983, **13**, 682.
[11] E. Küllik and M. Kaljurand, *Anal. Chim. Acta*, 1986, **181**, 51.
[12] R. S. Lehre and J. C. Robb, *J. Gas Chromatog.*, 1967, **5**, 89.
[13] J. Q. Walker, *J. Chromatog. Sci.*, 1977, **15**, 267.
[14] L. Levy, D. Fanter and C. Wolf, *Anal. Chem.*, 1972, **44**, 38.
[15] C. Bühler and W. Simon, *J. Chromatog. Sci.*, 1970, **8**, 323.
[16] W. J. Irwin, *Analytical Pyrolysis: A Comprehensive Guide*, Decker, New York, 1983.
[17] G. O. Piloyan, *Introduction to Theory of Thermal Analysis*, p. 67. Nauka, Moscow, 1964 (in Russian).
[18] S. Hiratsuka and A. Ichikawa, *Bull. Chem. Soc. Japan*, 1967, **40**, 2303.
[19] J. K. Foreman and P. B. Stockwell, *Automatic Chemical Analysis*, Horwood, Chichester, 1975.
[20] D. P. Carney and J. B. Phillips, *HRC&CC*, 1981, **4**, 413.
[21] J. R. Valentin, G. C. Carle and J. B. Phillips, *HRC&CC*, 1982, **35**, 269.
[22] K. Jinno, D. P. Carney and J. B. Phillips, *Anal. Chem.*, 1986, **58**, 1251.
[23] J. R. Valentin, G. C. Carle and J. B. Phillips, *HRC&CC*, 1983, **6**, 621.
[24] W. G. Laster, J. B. Pawliszyn and J. B. Phillips, *J. Chromatog. Sci.*, 1982, **20**, 278.
[25] D. P. Carney and J. B. Phillips, *Anal. Chem.*, 1986, **58**, 1251.
[26] V. V. Braznikov, Y. N. Bogoslovsky, E. P. Skorniakov, I. S. Zimin and V. N. Kurilenko, *USSR Certificate of Invention* No. 832472, *Bulletin Inventions*, No. 19 (1981) (in Russian).
[27] D. R. Deans, *J. Chromatog.*, 1984, **43**, 289.
[28] M. Kaljurand and E. Küllik, *Trends Anal. Chem.*, 1985, **4**, 200.
[29] D. J. Abbott, *HRC&CC*, 1984, **7**, 577.
[30] H. Brötell, G. Rictz, S. Sandqvist and M. Berg, *HRC&CC*, 1982, **5**, 596.
[31] D. Wright, *Int. Lab.*, 1985, **15**, Nov/Dec., 52.
[32] F. Müller and U. K. Goekeler, *Int. Lab.*, 1985, **15**, June, 42.
[33] *Chrompack News*, 1985, **12**, 1.
[34] A. W. Rechten, *Fluidik*, Springer Verlag, Berlin, 1976.
[35] G. Gaspar, P. Arpino and G. Guiochon, *J. Chromatog. Sci.*, 1977, **15**, 256.
[36] G. Gaspar, G. Vidal-Madjar and G. Guiochon, *Chromatographia*, 1982, **15**, 125.
[37] C. P. M. Schutjes, C. A. Cramers, C. Vidal-Madjar and G. Guiochon, *J. Chromatog.*, 1983, **279**, 269.
[38] R. Annino, M. F. Gonnord and G. Guiochon, *Anal. Chem.*, 1979, **51**, 379.
[39] R. Annino and J. Leone, *J. Chromatog. Sci.*, 1982, **20**, 19.
[40] *Valves for Chromatography*, Chrompack, 1978.
[41] C. H. Lochmüller, K. R. Lung and T. L. Lloyd, *Zymark Laboratory Automation Newsletter*, 1985, **2**, 6.
[42] C. H. Löchmuller, K. R. Lung and Cushman, *J. Chromatog. Sci.*, 1985, **23**, 429.
[43] C. G. Enke, *Science*, 1982, **215**, 785.

5

Application of multiple-input computerized chromatography

Applications of multiple-input chromatography are distributed over various branches of chemistry, though their number is not yet large. The absence of commercially available instruments is a restraining factor. Two-dimensional separation seems to be an exception. Thus, a number of challenges have been left to the investigators to invent new procedures, and to their enthusiasm in constructing new equipment by combining conventional instruments with specially designed units.

In this chapter we review in more detail two applications of multiple-input computerized chromatography in which significant results have been achieved, *viz.* evolved gas analysis (EGA) in thermal analysis, and determination of trace compound concentrations in environmental matrices. Although the selection of these applications is partly due to our own special interests, the main reason is that in the determination of the gaseous products of chemical reactions and in trace compound measurements, multiple-input chromatography has achieved more significant results than in other applications. In these applications we have evidence of the new information and experience that the multi-input technique may give.

5.1 CHARACTERIZATION OF GASES EVOLVED FROM POLYMERS

A wide variety of high molecular-weight materials can be found in nature and as products of technological processes. During the last two decades a large number of new polymers have been developed. One of their characteristics is marked heat resistance combined with good mechanical strength, making them useful in a great variety of engineering applications.

High molecular-weight substances of natural origin or those synthesized by man are generally unsuitable for direct analysis by modern instrumental techniques. The main difficulty is that they can be dissolved either not at all or only in selected solvents. This is the reason why in polymer analysis, indirect methods must very often be used. These problems arise not only in gas chromatography but also in ultraviolet (UV), infrared (IR) and nuclear magnetic resonance (NMR) spectrometry. Sometimes even special sample preparation techniques such as the KBr pellet technique in IR spectrometry are not applicable, because some polymers are difficult to grind to the required size, so that sample preparation needs a special technique and is time-consuming. In gas chromatography, the only difficulty arises when the sample to be analysed cannot be vaporized. The current chromatographic column techniques require that the vapour pressure of the compound under study be at least several hundred Pa at about 350°C, to enable the compound to pass through the column.

Thus it is understandable why numerous investigators have directed their efforts towards developing and introducing new methods for characterization and analysis of high molecular-weight compounds. In this case we can use an indirect approach in which the sample is converted into a state allowing its characterization. One of the methods widely used is thermal degradation of the sample followed by gas chromatographic analysis of the gaseous products evolved. Thermal degradation is also in use in organic synthesis as the simplest way of obtaining certain substances. A study of the thermal degradation products allows us to reach some conclusions about the original structure of the sample compound. The composition of the gases evolved during thermal degradation gives direct indications of the changes in polymers. The most important region of thermal degradation is that at moderate temperature, where the changes in the polymer are externally unnoticeable but may influence its mechanical properties. Besides thermal degradation, chemical and physical degradation methods are also in use, e.g. photolysis, radiolysis, hydrolysis, oxidation and reduction.

Figure 5.1 illustrates methods for generating evolved gases, and also some of the more common analytical techniques used. As can be seen from the scheme presented, the gas chromatograph is in a central position as a separator of mixtures as well as a basic independent analytical instrument. Most often, a combination of pyrolysis and gas chromatographic analysis has been used. When linked to a mass or an infrared spectrometer, the main task of the gas chromatograph is to separate the evolved gases for further identification. The application of Fourier transform infrared spectroscopy (FT–IR) to the direct analysis of evolved gases is increasing and in the future the combination of gas chromatography and FT–IR will become more dominant. A mass spectrometer is not available in some laboratories, owing to its high cost, and therefore the application of mass spectrometry is universal and may often be applied only when special or detailed studies are required. The main difference between pyrolysis gas chromatography and pyrolysis mass spectrometry is that in the latter the pyrolysis takes place in the ion source, i.e. under vacuum, and permits no secondary reactions between the primary fragments produced and thus gives information free from any side-effects. As seen from the scheme, the same technique can be used in desorption, catalytic and chemical reaction studies.

Another trend in polymer characterization is the use of methods of thermal analysis (e.g. differential thermal analysis, differential scanning calorimetry, thermogravimetry, thermomechanical and electrothermal analysis) in combination with evolved gas analysis. These techniques need a well-controlled furnace temperature and sensitive detectors to detect the gases evolved. The gaseous products obtained from polymers can be monitored by any gas detector that is based on

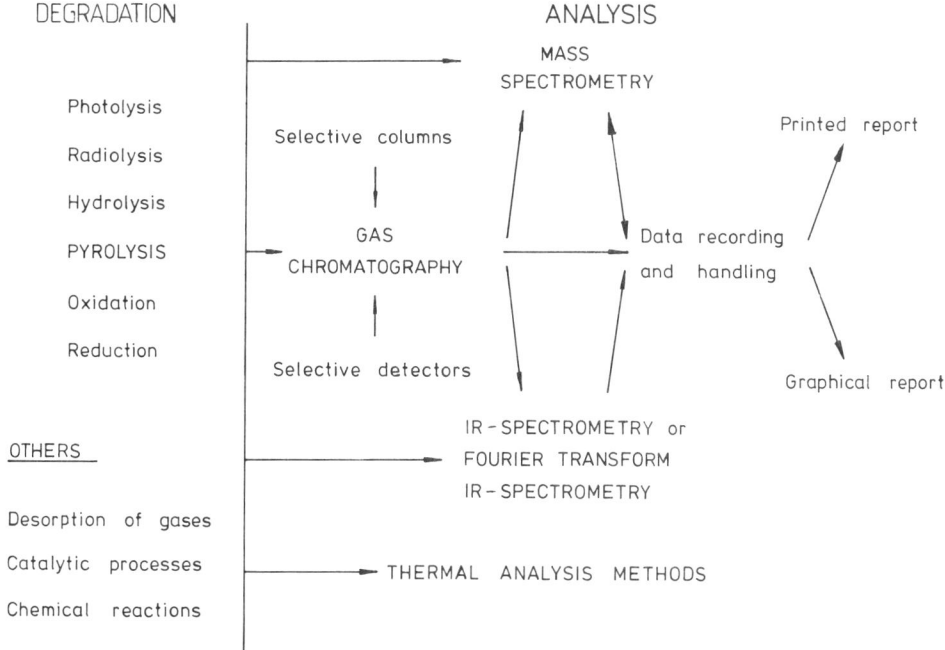

Fig. 5.1 — Methods of generating and analysing evolved gases.

thermal conductivity, flame ionization, pressure, density, volume, radioactivity or electrochemical and photometric behaviour.

All the degradation methods presented in Fig. 5.1 are in use but thermal degradation is the most common. The most convenient and inexpensive method of analysis is gas chromatography. As the aim of this chapter is to present multiple-input methods for use in gas chromatography, the other techniques are considered only briefly to emphasize the place of pyrolysis gas chromatography and thermochromatography among them.

Thermogravimetry and gas chromatography were introduced almost simultaneously. Although study of the possibilities of the thermogravimetric method and its development began as long ago as 1903, its serious introduction began only in the 1950s together with the development and introduction of gas chromatography and pyrolysis gas chromatography. The latter, particularly the thermal degradation, has deep roots in the progress of chemistry and therefore its development will be reviewed later.

Since 'classical' evolved gas analysis methods are one-dimensional, the multi-input methods enable us to add a new dimension to them. The first dimension characterizes the nature of the evolved products and the second, the temporal or temperature-dependent behaviour of the evolved products, allows calculation of the kinetics of gas evolution. The additional dimension remarkably expands the amount of information obtained from the thermal degradation of substances. The one-dimensional evolved-gas chromatographic methods are being successfully used for analytical purposes. We now have thousands of pyrolysis gas chromatograms, each usable as a 'fingerprint' of the corresponding compound. Unfortunately this use is rather limited because the chromatograms are obtained under different conditions in different laboratories. The two-dimensional multi-input methods have an analytical value but in addition, the kinetic information available from the two-dimensional output can lead to a better understanding of the degradation process. Therefore, we are now only at the beginning of understanding the possibilities of multi-input methods. The evolving gases give some indication of the structural changes in the polymers that improve their thermostability or decrease their flammability. The potential toxicity of the evolved products (hazardous gases) is another important property to be taken into account. This list could be continued by every polymer chemist. Knowing the evolution kinetics enables us to solve several important problems in polymer chemistry and technology.

5.1.1 Various approaches to polymer characterization

5.1.1.1 Pyrolysis mass spectrometry

Pyrolysis combined with mass spectrometry is a powerful analytical tool for the analysis of evolved gaseous products of thermal degradation and is widely used in polymer chemistry. In studies of pyrolysis products by mass spectrometry we again meet a phenomenon where processes that are the same in principle are given different names, depending on whether the user is a gas chromatographer or a thermal analysis researcher. The chromatographers use the term 'pyrolysis mass spectrometry' and the others use 'evolved gas analysis by mass spectrometry'. In this book we consider the problem from the point of view of chromatography and use the terms of gas chromatography. The early developments in the topic have been presented by Langer [1]. Generally two analytical schemes are in use. First, pyrolysis mass spectrometry (Py–MS) where thermal degradation products are directed to the mass spectrometer without prior separation into individual components, and secondly pyrolysis gas chromatography–mass spectrometry (Py–GC–MS). Pyrolysis mass spectrometry was the first of the integrated pyrolysis techniques to be described. Obviously the development of gas chromatography and particularly pyrolysis gas chromatography restricted, to some extent, the mass spectrometer applications to polymer studies. It is understandable because pyrolysis gas chromatography proved to be less expensive and need less complicated instrumentation than did pyrolysis mass spectrometry.

Mass spectrometry as an analytical method is very sensitive and gives a large amount of information but at the same time is difficult to handle. It allows the use of several different methods of ionization, such as the electron impact, chemical, electric field and desorption modes. The availability of on-line computer data-handling systems initiated reappraisal of pyrolysis mass spectrometry.

In some sense, pyrolysis mass spectrometry and pyrolysis gas chromatography are comparable in the analysis of molecular-weight compounds. Both need prior thermal decomposition of the substance and the results are obtained as a single gas chromatogram or a single mass spectrum, respectively. In contrast to pyrolysis gas

Sec. 5.1]   **Characterization of gases evolved from polymers**   143

chromatography, in which the pyrolysis takes place in a carrier gas atmosphere (He, Ar or N_2), pyrolysis in the mass spectrometer takes place in a vacuum, i.e. under conditions similar to those in extraterrestrial space, so secondary reactions between the degradation products are avoided.

The possibilities of combining pyrolysis and mass spectrometry in high molecular-weight compound studies are shown in Fig. 5.2. Direct-probe pyrolysis mass

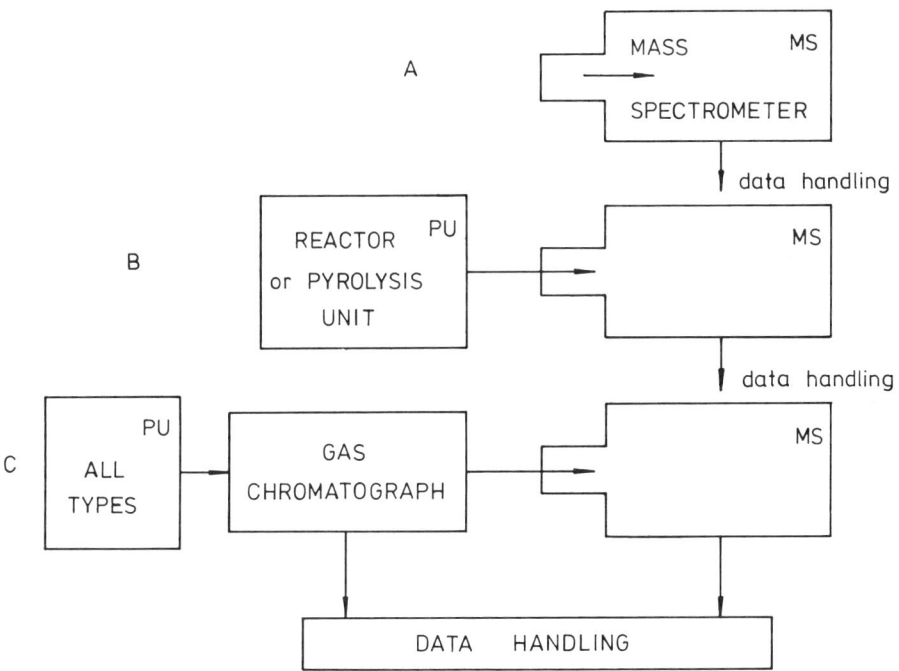

Fig. 5.2 — Mass spectrometry of evolved gases: A, pyrolysis in an ion source; B, pyrolysis unit coupled with a mass spectrometer; C, separation of evolved gases by using gas chromatography before mass spectrometric analysis.

spectrometry (sample degradation in the ion source) has been in use from the start of this technique (Fig. 5.2A). The temperature of the sample is raised either by using the ion chamber heater or by a solid sample inlet system. However, pyrolysis directly in the ion source is not generally recommended, because of possible contamination. In most cases, the characteristic features of direct sample systems are a slow heating rate and a relatively long residence time of the products in the degradation zone. However, the direct sample technique is very handy, is available in most mass spectrometers and needs no special pyrolyser.

The second possibility is that the pyrolysis unit is an independent device and is easily connected to the mass spectrometer (Fig. 5.2B). This version permits the use of different types of pyrolysis system, such as the Curie-point, filament or laser pyrolysis devices. For maximum transfer efficiency, the pyrolysis should be effected as near as possible to the ion source of the mass spectrometer. There is still the possibility of contamination of the ion source and the connecting tubing, which could influence the results. Without a separation of the mixture of degradation products the mass spectrum is the sum for all the components formed during the thermal degradation.

The third possibility for pyrolysis mass spectrometry is that the pyrolysis products are swept into the gas chromatographic column and the separated components are, in turn, eluted into the ion source of a rapid-scan mass spectrometer. The inclusion of the pyrolysis unit in the combined gas chromatograph–mass spectrometer system provides decomposition of sample, separation of the degradation products and their qualitative detection by means of a single integrated operation (Fig. 5.2C). The whole system is quite complicated and to obtain the best results, careful optimization of all three independent units is needed.

From the point of view of characterizing evolved gases, the second is the most widely used system. This system is more flexible because there are several possibilities of expanding the analytical determination.

The practical applications of Curie-point pyrolysis mass spectrometry have been summarized by Meuzelaar in the *Compendium and Atlas of Mass Spectra of Recent and Fossil Biomaterials* [2]. The spectrum atlas has been divided into eight sample groups: (1) carbohydrates and glyco-conjugates, (2) peptides and proteins, (3) nucleotides and nucleic acids, (4) lipids, (5) natural products, (6) humic materials and geopolymers, (7) biochemicals (drugs, vitamins, etc.), (8) polymers of non-biological origin (plastics, resins, etc.). Among the biomaterials, synthetic polymers were included on the grounds that they may be present as contaminants in biochemical samples or may serve as model compounds for some biomaterials.

The compounds under study were coated on the Curie-point wire by application as fine suspensions in methanol, which seems to be an inert solvent producing only a small residual fragment peak. The solution or suspension concentration was in the range 0.1–0.3%. Generally, the solvent was evaporated from the wire at a constant speed of rotation of the wire and at reduced pressure. The optimized analysis conditions used in pyrolysis mass spectrometry by Meuzelaar may be taken as a model, and are therefore given in Table 5.1 for reference.

The above-mentioned pyrolysis–mass spectrometric technique does not offer calculation of the evolution kinetics, because the temperature of the inlet port is usually kept constant. The advantage of mass spectrometry is the high detection speed, allowing a large number of scans to be made during a pyrolysis. In one sense, one scan in mass spectrometry is equal to a single injection in gas chromatography, and therefore multiple scanning in mass spectrometry can be compared with multiple input in gas chromatography. However, the sample heating rate should be standardized with a linear temperature programme. Thermal degradation mass spectrometric analysis was developed by Risby and Yergey [3]. They investigated the thermal degradation of polystyrene and poly(vinyl chloride). The temperature-programmed system and controller were used to allow the temperature of the heating element to be increased linearly or stepwise over the temperature range chosen. The main idea of temperature control was to use the pyrolyser filament (a

Table 5.1 — Standard experimental conditions in pyrolysis mass spectrometry. (Reproduced from [2] by permission. Copyright 1982, Elsevier Scientific Publishers)

Experiment parameter	Selected value or mode
wire cleaning	reductive heating
equilibrium temperature	510°C
temperature rise-time	0.1 sec
total heating time	0.9 sec
inlet temperature	150°C
electron energy	14 eV
scanning speed	10 spectra/sec
mass range	m/z 15–162
number of scans accumulated	150

platinum wire) as a sample heater and also as a resistance thermometer. The average temperature over the platinum wire was calculated from its resistance and was controlled by an on-line computer. At each temperature step, a mass spectrum was collected, until thermal degradation was complete. The rate of temperature increase used depended on the nature of the substances under study. The possible rates were 0.05–300°C/sec. To simplify the data handling, some variables were kept constant during the experiments, such as reactant gas (methane) pressure 1 mmHg, source temperature 120°C, 20 mass spectral scans per minute and with recording over the 80–480 m/z range for polystyrene, and 60–460 m/z for poly(vinyl chloride) (positive ion detection) and 10–410 m/z (negative ion detection). The first ten mass spectra were collected before starting to heat the platinum filament. This 30-sec delay was found to increase the reproducibility of the data. At the end of that time the temperature had risen to 70°C as a result of heat conduction from the source. After this initial hold time the voltage to the platinum filament was increased stepwise until the temperature was 540°C, at which it was held for 30 sec to ensure total removal of the sample. The rate of temperature increase varied from 1 to 16°C/sec.

The results of time-resolved pyrolysis of uncross-linked polystyrene (MW=10^5) are presented in Fig. 5.3. A two-dimensional plot was constructed as intensity *vs.* m/z and temperature. Here the multiple-input technique allows construction of additional data presentations as in Fig. 5.4, in which the two-dimensional plot is used to build a specific ion profile for the thermal degradation. The data obtained were used to calculate an activation energy.

Pyrolysis mass spectroscopy seems to be a more powerful and rapid technique than pyrolysis gas chromatography, for evolved gas analysis. However, the techniques complement each other. A detailed analysis shows that the scan time is not remarkably longer in mass spectroscopy than in high-speed chromatography (about 0.25–1 sec [4]). As already pointed out, the degradation conditions in mass spectrometry are comparable to those in space. In pyrolysis gas chromatography, it is always possible to degrade samples in different atmospheres (e.g. oxygen, nitrogen, air and inert gases) at different pressures. In pyrolysis mass spectrometry, it is difficult to use different environments. The cost of pyrolysis gas chromatographic equipment is about two orders of magnitude less than that for pyrolysis mass spectrometry. Last but not least, the mass spectra of some compounds are quite similar, whereas their pyrograms are rather different. As will be demonstrated in Sections 5.1.2 and 5.1.3,

146 Application of multiple-input computerized chromatography [Ch. 5

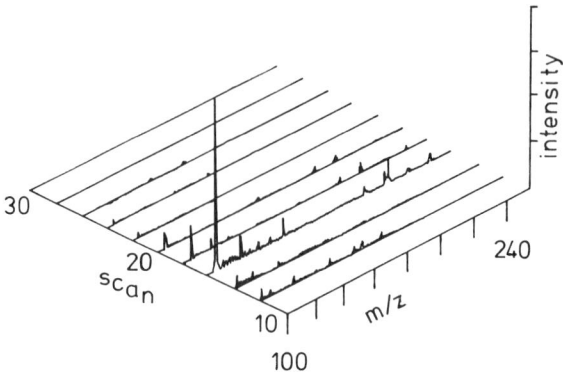

Fig. 5.3 — Three-dimensional plot for thermal degradation of uncross-linked polystyrene (by heating at 16°C/sec), shown as intensity *vs.* scan number and *m/z*. (Reprinted with permission from T. H. Risby, J. A. Yergey and J. J. Scocca, *Anal. Chem.*, 1982, **54**, 2228. Copyright 1982, American Chemical Society).

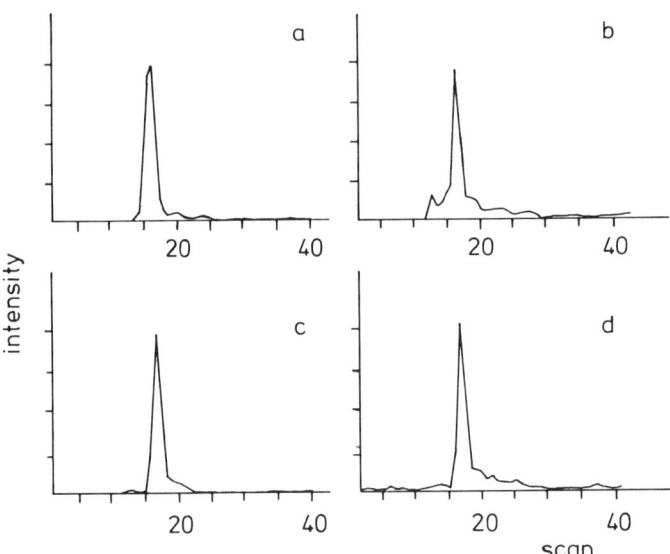

Fig. 5.4 — Specific ion profiles for thermal degradation (at 16°C/sec heating rate) of uncross-linked polystyrene, shown as intensity *vs.* scan number: (A) *m/z* 105, temperature of evolution maximum (TEM) 423°C; B, *m/z* 117, TEM 423°C; C, *m/z* 209, TEM 423°C; (D), *m/z* 221, TEM 423°C. (Reprinted with permission from T. H. Risby, J. A. Yergey and J. J. Scocca, *Anal. Chem.*, 1982, **54**, 2228. Copyright 1982, American Chemical Society).

Sec. 5.1] Characterization of gases evolved from polymers

pyrolysis gas chromatography represents an independent analytical method and not a 'poor-man's spectrometry', and has its own problems and their solutions.

A complete review of pyrolysis–mass spectrometric techniques has been given in the excellent books by Irwin [5] and Hmelnitsky et al. [6].

5.1.1.2 Pyrolysis infrared spectrometry

Measurements of the absorption spectra of gas chromatographic effluent peaks have been used and discussed but until now IR spectrometry has not been widely used to analyse gases evolved in thermal degradation, because of the relatively low sensitivity of the instrument and the low concentration of most components obtained by conventional pyrolysis gas chromatography. The possibilities of using IR spectrometry in evolved gas analysis are shown in Fig. 5.5. Recent developments of this

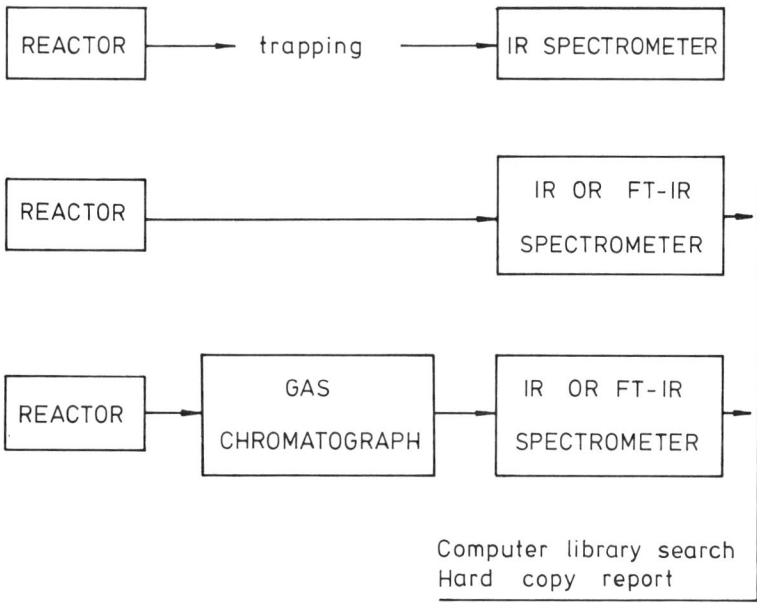

Fig. 5.5 — IR spectrometry of evolved gases.

technique have been directed towards improving the detection sensitivity. The situation changed with the advent of Fourier transform infrared (FT–IR).

Despite there being the possibility of mating the separating power of the gas chromatograph with the identifying capacity of an infrared spectrophotometer, it is well known that their coupling and application as a united system is somewhat complicated. These problems arise from the basic operating mismatch between gas chromatography and infrared spectrometry. Several possibilities of using a gas chromatograph and an infrared spectrometer for evolved gas analysis have been suggested. One possibility is to trap the separated compounds at the gas chromatograph outlet and introduce them later into an infrared spectrometric cell for spectral analysis. This technique is time-consuming and often unsuccessful because trapping may be accompanied by sample loss (during sample adsorption–desorption). Also, trapping needs a special technique because the amounts of components separated in an analytical gas chromatograph are usually below the detection limit for a conventional IR spectrometer. In conventional gas chromatography, the optimum sample size is several orders of magnitude smaller than that for IR spectrometry. A standard dispersive infrared spectrometer requires a relatively large amount of sample compared with that needed for gas chromatography. The other problem is the matching of carrier gas velocity and IR spectrometer scanning speed because the separated components appear from the gas chromatograph much faster than they can be scanned by the infrared spectrometer. A conventional IR spectrometer has a scanning time of about 3–10 min per spectrum, which is considerably longer than the time taken for elution of a band from a chromatographic column.

Recently, a modern ratio-recording infrared spectrometer was used successfully to characterize polymeric materials by analysis of the gases evolved from thermally degraded samples [7]. With a temperature rise of 5°C/min, a scan time of 1–2 min was found to be sufficient to define the system, i.e. a dispersive instrument with fast scanning capability may be used for evolved gas analysis. Since the evolved gases do not contain many components, there are usually individual or easily corrected IR bands available for constructing the evolution profiles. Both the continuous and stopped flow modes can be used.

Typical spectra of gases evolved from continuous flow pyrolysis of polyether urethane at different temperatures with different recording sensitivities are given in Fig. 5.6 [7]. As can be seen, the spectra are not so clearly characteristic as chromatograms in pyrolysis gas chromatography and at higher sensitivities the interpretation is complicated.

For computerized data-handling the peak maximum retention time in GC is more useful than the absorption maximum in IR spectra. It has been pointed out [7] that for accurate and reproducible construction of the evolution profiles the temperature corresponding to the absorption band measurement must be determined for each spectrum. It was recorded at the starting time of the first, last and intermediate scans, along with the number of scans, the heating rate and the starting time of the experiment. After that the absorbance and temperature for each band of interest were calculated. Data handling was somewhat complicated because the acquisition program cycle time was not constant, as the time needed to record the files in the computer memory is slightly increased with each successive record. Using dispersive IR spectrophotometers at a high scan-rate may result in a relatively poor signal-to-noise ratio and digitization rates that do not allow extensive data smoothing. Thus several problems arise in using the technique, but the evolution profiles for some thermal degradation products from cross-linked poly(methyl methacrylate) give information about their thermal degradation in the temperature range 100–450°C (Fig. 5.7). On the other hand, the information about thermal degradation is not complete, because only the monomer, carbon dioxide and methanol were taken into account.

In conclusion, the dispersive spectrometer slowly scans and records the concen-

Sec. 5.1]   **Characterization of gases evolved from polymers**   149

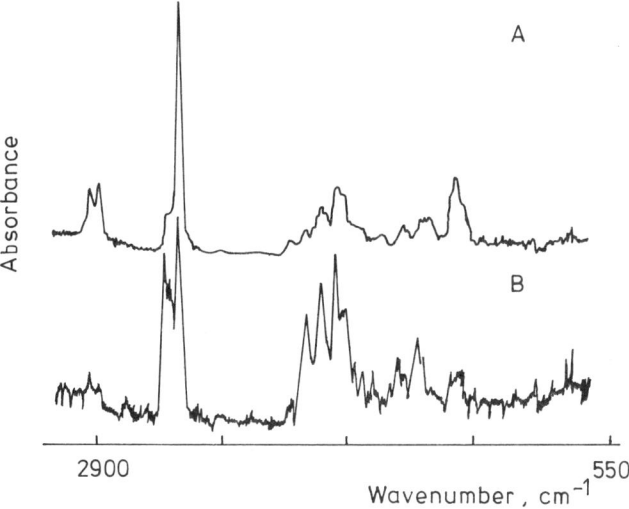

Fig. 5.6 — Spectra of gases evolved from continuous flow pyrolysis of polyether urethane at 300°C (×10) (A) and 330°C (×4) (B). (Reprinted with permission from R. G. Davidson and G. I. Mathys, *Anal. Chem.*, 1986, **58**, 837. Copyright 1986, American Chemical Society).

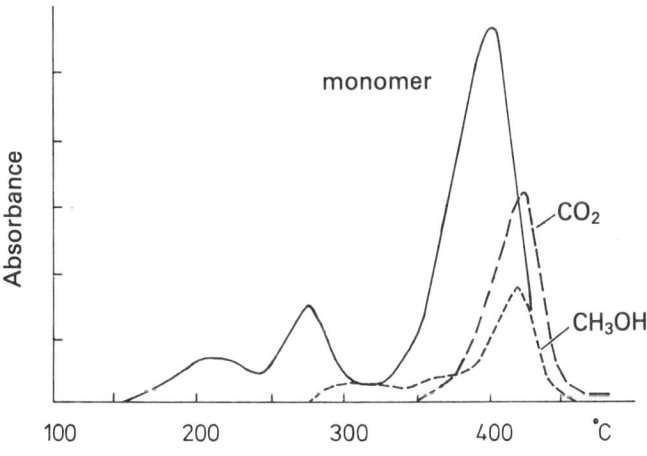

Fig. 5.7 — Evolution profiles of cross-linked poly(methyl methacrylate) PLEX 55 Temperature rise 5°C/min. Bands: monomer (1310 cm^{-1}), CO_2 (2360 cm^{-1}) and methanol (1050 cm^{-4}). (Reprinted with permission from R. G. Davidson and G. I. Mathys, *Anal. Chem.*, 1986, **58**, 837. Copyright 1986, American Chemical Society).

tration changes in the cell and the spectral subtractions can only be performed on a short section of spectrum and even then with caution. The gas chromatographic separation time in the column in thermochromatography (a high-speed system) and the scan time in IR spectroscopy are in the same range. In gas chromatography, the results depend on the separation ability of the column and in most cases, there is always hope of finding more suitable conditions for the perfect separation of the evolved gases.

The third approach is the use of on-line GC–FT–IR systems. These spectrometers are capable of producing spectra with a very high signal-to-noise ratio and a large dynamic range. A complete infrared spectrum can be recorded in less than a second. The gas chromatograph infrared accessory can be used with almost any modern gas chromatograph that already has a flame-ionization or thermal conductivity detector. The combination of gas chromatographic FT–IR and chromatographic mass spectrometric analysis techniques can provide better identification of components and even of isomers.

FT–IR spectrometry offers new possibilities in evolved gas analysis because the scanning is 10–1000 times as fast as that in dispersive instruments, allowing identification of gas chromatographically separated peaks. As the IR spectra libraries contain thousands of spectra, identification becomes easier. However, despite the recent achievements of GC–FT–IR no references are available to the authors where this valuable technique has been applied in evolved gas analysis. The FT–IR spectrometer is rather an expensive analytical instrument and not universally available.

5.1.1.3 Methods of thermal analysis

Methods of thermal analysis have been widely applied to substance characterization [8]. Although they mainly treat the sample itself and not the degradation products, it is useful to have a brief look at these methods to understand better the role of evolved gas analysis. The most frequently used techniques in thermal analysis are as follows.

(1) Differential scanning calorimetry: this measures the difference in the rate of energy flow to (or from) the sample and a reference material as a function of temperature or time.
(2) Differential thermal analysis: this measures the temperature difference between the sample and the reference material as a function of temperature when both are subjected to a controlled temperature programme.
(3) Thermomechanical analysis: this measures the dimension or deformation of a substance under a non-oscillatory load, as a function of temperature.
(4) Thermogravimetry: this is defined as a technique in which a specimen is heated in a controlled atmosphere and the specimen is monitored as a function of temperature or time. Thermogravimetry allows measurement of material properties such as thermal stability, softening temperature, holding points, phase diagrams, expansion and flow behaviour, heat capacity, heat of transition and reaction, purity, curing reactions, glass transition, impact resistance and reaction kinetics. The weight-loss curve recorded during heating is characteristic for each substance. Besides thermogravimetry, there is a much older gravimetric method based on weighing. In gravimetry, the sample is weighed after extended periods of isothermal heating.

To perform a thermogravimetric experiment the analyser must be capable of performing heating and weighing simultaneously. A thermogravimetric analyser includes a high-sensitivity recording microbalance, a furnace, a sample holder, a temperature programmer, a pneumatic gas system for purging the furnace and sample chamber, gas switches (sometimes equipped with facilities for evolved gas analysis) and a data-acquisition system. The instrument used for thermogravimetric measurements is also referred to as a thermobalance.

As in all methods of thermal analysis, attention must be paid to calibration and determination of the heating temperature. Very often, the thermocouple is not in direct contact with the sample and that makes the determination of the exact temperature more difficult.

Thermogravimetry and pyrolysis gas chromatography (PyGC) are, in a sense, equivalent methods because the chemical nature of the basic process is the same. The difference lies in the nature of the output signal. The weight-loss curve is a characteristic of the sample and reflects the individual contribution of each component evolved to the overall response. In PyGC, gases evolved during heating are separated in a chromatographic column and the peak amplitude and retention time characterize the degradation process. Owing to the separation of the mixture into individual components the amount of information in PyGC is far higher than in thermogravimetry.

For some special purposes, thermogravimetry, gas chromatography [9] and mass spectrometry [10] are coupled to investigate thermal decomposition.

5.1.1.4 *Historical remarks: development of the pyrolysis method*

To understand why thermal degradation and in particular, pyrolysis gas chromatography, are so widely used for the characterization of high molecular-weight substances, it is beneficial to know the background of this method. The pyrolysis method dates back to the early days of chemistry. It is remarkable that in the development of chemistry and chemical technology, fire and heat have been of great importance. Thermal degradation and then distillation of the organic or mineral material was one of the few methods available to the alchemist and later to the chemist to obtain new and purer substances.

Thermal degradation or pyrolysis (Greek: pyr, fire; lysis, dissolution) is the conversion of one substance into another substance or several substances by the agency of heat alone. The final result of this process is a set of molecules of lower mass or, depending on the degradation conditions, products with an increased molecular weight through intermolecular interaction.

The change in substances at high temperatures became a well-recognized principle which was emphasized by events marking a turning point in chemistry, such as the formation of sulphuric acid by heating ferrous sulphate. This event was described by Basil Valentine at the end of the fifteenth century. The element phosphorus was discovered in 1669 by Brandt, who destructively distilled the residues separated from urea. In 1825, Faraday isolated benzene and in 1834, Mitscherlich prepared benzene by vigorous distillation of benzoic acid. Bertholet was the first to prepare biphenyl from benzene. In this case, the final product has a higher molecular weight than the initial substance. This kind of change is now brought about by the use of modern techniques. The vapour of organic compounds is passed through a red-hot porcelain or iron tube. The reaction tube temperature is chosen according to the initial substance and results expected. The pyrolysis of organic compounds attracted much attention in the first decades of the twentieth century but lost importance when the emphasis in organic chemistry shifted towards the study of more complicated methods of synthesis.

In analytical practice, the larger naturally occurring molecules were extracted and purified only with considerable effort and were not usually suitable for either thermal degradation or the only other method available at that time, 'hot-tube' chemistry. Caoutchouc was the first polymer to be thermally degraded, in the nineteenth century. Approximately a century later, in 1935, the first synthetic polymer, polystyrene, was studied by Staudinger. The product of thermodegradation was a complex mixture of hydrocarbons, including monomers, dimers, trimeric and even tetrameric sub-units, depending on the degradation pressure, temperature, and heating rate. It was also shown that the monomer yield was dependent on the sample weight. Up to that time, the main aim was to explain or to use thermal degradation to restructure macromolecular compounds for industrial purposes. Applied pyrolysis became common in petrochemistry in treating organic geo-polymers such as coal, oil shale, and kerosene, or biopolymers such as wood.

The large-scale application of pyrolysis in chemical technology needed more effective analytical methods for determination of thermal degradation products and this was the reason for the acceleration in the development of analytical pyrolysis. Pyrolysis is used to aid in the analysis of the high molecular-weight compounds which are difficult to study by more conventional methods. The concept of the free-radical chain reaction [11] as a model of thermal reactivity was one of the aspects of the study of the theory of pyrolysis.

Of great importance in the development of analytical pyrolysis was a combination of pyrolysis with a sophisticated separation technique and identification of the fragments evolved. A developing technique at that time was mass spectrometry, which was and still is an effective method for characterizing the fragments or individual pyrolysis compounds. The identification of fragments was aided by the recognition that polymer pyrolysis yielded hydrocarbon fragments and mass spectrometry was a useful technique for the analysis of hydrocarbon mixtures. The first practical examples were thermal decomposition of polystyrene, performed by Madorsky and Straus [12] and later, some vinyl polymers by Wall [13]. Madorsky and Straus studied the mechanism of thermal degradation of polymers and showed that the type and relative amount of pyrolysis products are a function of the molecular structure, and type and number of side-groups. They also showed that thermal stability and the degradation products obtained by pyrolysis may be related to the strength of the C–C bonds in the polymer chain. This means that the degradation products and their quantity are a function of the chemical composition and molecular order in the substrate as well as of the degradation conditions.

Besides mass spectrometry, infrared spectrometry was also used. In that case, the main goal was to identify polymers. This technique was useful for characterizing polymeric materials capable of being made into thin films. The properties of many polymers deny this approach and for polymer characterization by IR spectrometry a special method of sample handling was needed. One of the first examples was presented by Barnes *et al.* [14] who studied the mixture of pyrolysis products from a highly vulcanized rubber. After the introduction of gas–liquid chromatography by Martin and James [15] in 1952, mass application of IR spectrometry to degradation studies slowed down. A renaissance began when a mass spectrometer was connected to a gas chromatograph (chromato-mass spectrometer) and Fourier transform–IR spectrometry (FT–IR) showed its power as an analytical tool when coupled with a gas chromatograph.

Soon after its introduction, gas chromatography was used for analysis of pyrolysis products. In 1954, Davidson *et al.* [16] showed that the gas chromatograms of pyrolysis products are specific for each substance. They suggested that the chromatograms of pyrolysis products, pyrograms, should be used to characterize a given polymer. In 1959, several papers were published on polymer identification and on the possibilities of combined off- or on-line pyrolysis gas chromatography [17]. Lehre and Robb [18] considered the possibilities of quantitative estimation of a polymer mixture or a copolymer composition. The samples studied were poly(vinyl chloride)--poly(vinyl acetate) copolymers. Of the gases evolved, the peaks of HCl and CH_3COOH were used to estimate the relative proportions of each monomer in the copolymer. At the same time, the pyrolysis products were analysed by IR spectrometry. The results obtained by both methods were promising and extremely good. Simplicity and accessibility were the main reasons why the use of pyrolysis gas chromatography and its combination with other methods of instrumental analysis increased.

Two tendencies in the development of analytical pyrolysis from the beginning of the 1960s should be mentioned. One was to find and understand the process and design an efficient and reproducible system. The other was concentrated on expanding the application of the technique and on showing its widsespread utility in many diverse fields. The latest thorough review of pyrolysis gas chromatography as an analytical method was given by Irwin [5]. This book is extremely valuable for the reason that the author considers not only pyrolysis methods and pyrolysis gas chromatography, but also reports on other techniques (e.g. pyrolysis mass spectrometry, data handling and taxonomy). The application of the method to the analysis of synthetic polymers, biological molecules, geopolymers and other high molecular-weight substances is also discussed. A detailed review of the pyrolysis gas chromatography papers in the above-mentioned areas is given. The author index contains more than 1400 names from 1300 papers, showing the extensive use of pyrolysis gas chromatography for the analysis of non-volatile materials and characterization of high molecular-weight substances.

5.1.2 Analytical and applied pyrolysis gas chromatography

Pyrolysis gas chromatography is an analytical method which produces a chromatogram of the thermal degradation products of a sample for sample identification or kinetic studies. In principle, pyrolysis gas chromatography may be considered to consist of two separate processes taking place in two independent units; a pyrolyser and a gas chromatograph, even when the pyrolysis unit and the analyser are connected in an on-line mode. The success of the analysis depends on the optimization of the degradation and analysis conditions.

In some special cases, separate pyrolysis in a closed vessel is used (Fig. 5.8A). This technique, the oldest in this field, is very simple but it gives no real picture of the

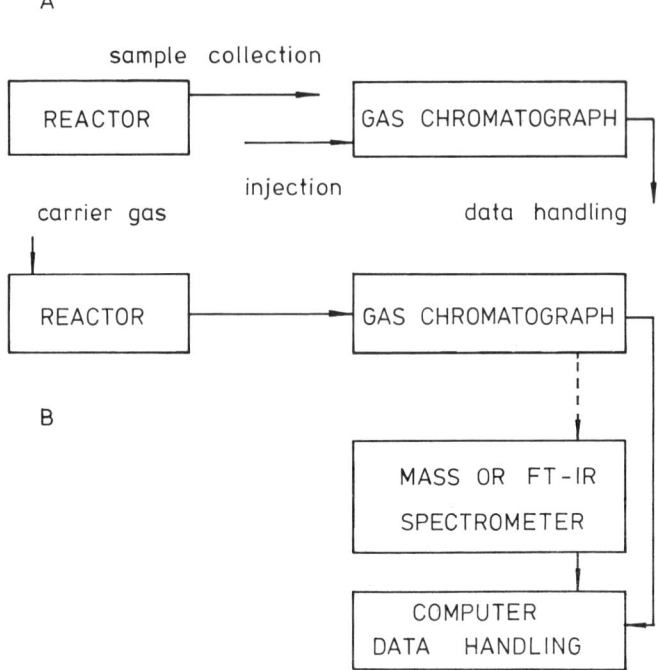

Fig. 5.8 — Pyrolysis and gas chromatography of evolved gases: A, separate pyrolysis and analysis; B, on-line system.

pyrolysis process because in the closed vessel, the pyrolysis products include not only primary but also secondary decomposition products. During the decomposition process, radicals are formed which may react or decompose to give new products. The scheme was in use in the early days of pyrolysis gas chromatography, when gas chromatography had not yet found its place as a separation method, for further analysis with a mass spectrometer or IR spectrometer. From an analytical point of view, the closed-system pyrolysis permits repeated analysis of the same sample and establishment of optimum analysis conditions. After a short period, the system was modified into the on-line mode (Fig. 8.5B). In this scheme, the pyrolyser is interfaced to the gas chromatograph injection port. In the analysis mode the carrier gas flows through the pyrolyser so that the volatile pyrolysis products are swept away into the analytical column. The carrier gas flow is continuous except during the insertion of a blank sample, and the carrier gas flow then by-passes the pyrolyser. This system is most commonly used in analytical pyrolysis. The advantages of the system are flexibility and the possibility of using other analytical devices.

The most flexible system is that shown in Fig. 5.9. It allows the degradation of a sample in an atmosphere of a selected gas and the regulation of the flow of relevant gases into an analytical column at suitable intervals. During pyrolysis in the first step the evolved gases are vented through a sample loop to the atmosphere. In the second

Sec. 5.1] Characterization of gases evolved from polymers

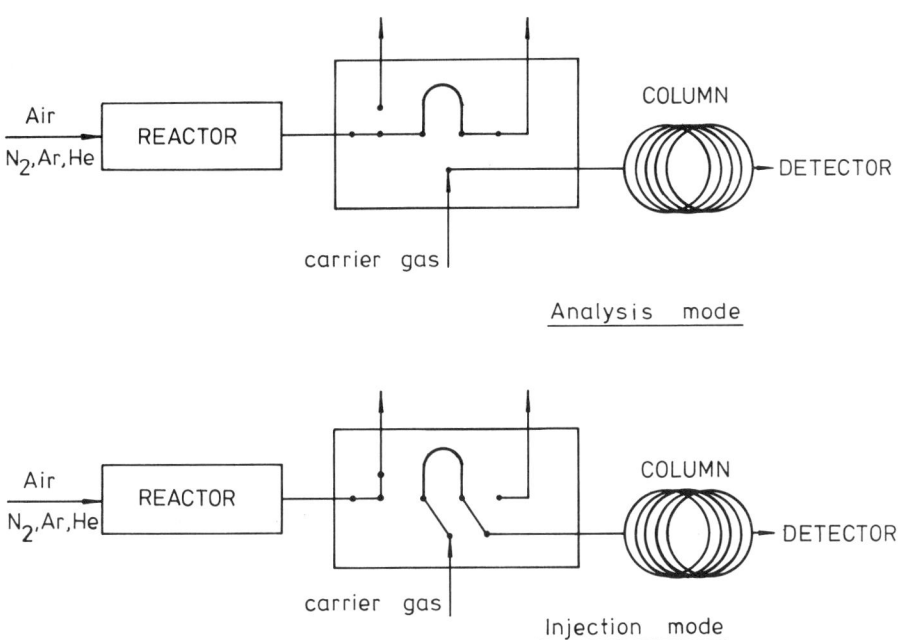

Fig. 5.9 — Universal scheme for studying evolved gases.

step the loop content is vented into the column and then the loop is switched back to the first mode. The interval between sampling depends on the separation time of the sample components. This scheme is universal for multiple-input gas chromatography and thermochromatography. The pyrolysis unit can be changed to a reaction vessel if a chemical reaction or a catalytic process is of interest.

Pyrolysis gas chromatography is easier to use than mass spectrometry and is also not so expensive. A great number of gas chromatographs are available in analytical and research laboratories and this makes it easy to combine chromatographs with additional pyrolysis units. Early pyrolysis units were home-made, but pyrolysis systems are now commercially available, including some for large scale studies. Experience in gas chromatography makes it easy to combine and use pyrolysis gas chromatography in polymer analysis. The drawbacks of the method are the following. As the degradation process is highly complicated, the chromatogram of pyrolysis products may contain hundreds of individual components all of which may give different responses (peaks). For that reason a peak area does not always correspond to the actual amount of the component present and a correction for each individual component (peak) is needed. Also, incomplete separation of peaks can cause errors in PyGC.

The popularity and usefulness of pyrolysis gas chromatography arises from its performance in the study of non-volatile materials, the most important of which are (1) natural and synthetic polymers and fibres (microstructure determination and quantitative analysis of copolymers, fingerprint identification etc.); (2) thermostable polymers; (3) resins and waxes; (4) biomolecules; (5) biopolymers (lipids, nucleic acids, proteins, carbohydrates); (6) drugs; (7) food; (8) fossil fuels (charcoal, kerosene, oil shale, tar sand); (9) soil, sediments, geochemicals; (10) microorganisms (bacteria, fungi), microbiological samples; (11) flame retardants and (12) organometallic compounds.

Pyrolysis gas chromatography has applications in: (1) forensic studies, pathology; (2) environmental studies, waste reclamation; (3) control of pyrolytic reaction products in organic synthesis; (4) catalytic processes and catalyst studies; (5) reaction kinetics and (6) detection of life in space, etc.

Pyrolysis techniques are divided into applied and analytical. Applied pyrolysis is mostly a large-scale preparative operation with an important industrial or semi-industrial application. A great number of special pyrolytic reactions are in use in organic chemistry. The most common techniques used are flow and flash vacuum pyrolysis. Flash pyrolytic methods, in particular, have permitted straightforward synthesis of many sensitive and reactive organic compounds which are not readily obtained by reaction in solution. This technique is often simple and requires only inexpensive apparatus for preparative scale synthesis. Analytical pyrolysis is a small-scale method having great importance in the analysis of relatively high molecular-weight compounds which are difficult to study in a more conventional way.

5.1.2.1 Single-input pyrolysis gas chromatography: conventional method

In single-input pyrolysis gas chromatography the sample is entirely degraded with a strong (low or high temperature) pulse and the degradation products are directed into the gas chromatograph column. The degradation is performed in some of the reactors described in Section 4.2.2. A single-input pyrolysis gas chromatographic operation is the same as in conventional GC, the only difference being that the injection of a gas mixture is replaced by that of the thermal degradation products formed in a special unit (a pyrolyser). This technique is one-dimensional and gives a signal in the form of a sequence of peaks which we call a chromatogram.

If we are able to guarantee that the conditions in the pyrolysis unit and those in the gas chromatograph will not vary, then we can expect reproducible pyrolysis gas chromatograms. Practical and theoretical studies show why it is difficult to achieve good reproducibility. It was shown in Section 4.2.2.2 that fluctuations in the pyrolysis temperature will produce variation in the composition of the gases evolved.

As well as single-input systems multiple-input systems have appeared and it is reasonable to call the first system conventional PyGC, which is analogous to conventional (single-input) gas chromatography. Single-input pyrolysis gas chromatography is a well developed technique and is used by a great number of investigators. This is due to its ease of use and relatively low cost. The pyrolysers which are now commercially available are convenient to handle and meet the needs of chromatographers in qualitative and quantitative analysis. While good reproducibility may be achieved in a given laboratory, interlaboratory reproducibility remains a problem even today. If the aim is reproducibility of pyrolysis and evolved gas analysis then efforts should be made to develop a more effective and reproducible system. The problem of finding the optimum conditions for pyrolysis, thermal degradation product separation, detection, identification and quantification is not an easy one. In practice it means that attention must be paid to the column (support and type), the temperature programming of the gas chromatographic oven, the detector sensitivity, the carrier gas velocity and other parameters that affect the analysis.

Single-input pyrolysis gas chromatography has found its place in the analytical chemistry of polymers. In scientific papers we can find a great number of pyrolysis chromatograms obtained under different conditions. Unfortunately, there are no catalogues of programs such as there are for infrared spectrometry (e.g. [19]). Since pyrolysis gas chromatography is an empirical method with very little theory developed, all sorts of chemometric methods are used in the classification and simplification of pyrograms, as well as in pyrogram data banks. Both gas chromatography-based pyrograms and pyrolysis mass spectra have been the objects of extensive chemometric studies. Although interesting and promising, chemometrics is beyond the scope of this book.

The mechanism of pyrolysis product formation is very complicated and up to now, no papers have been reported that give a complete determination of all the individual components of the evolved gases. As long as all the components of the evolved gases have not been identified, there is no sense in trying to establish the mechanism of thermal degradation. Taking into account that each substance gives a mixture of products that is characteristic of a given polymer, it needs considerable effort to identify all the degradation products. Even then a theoretical description of the degradation kinetics is rather complicated, and possibly hopeless. A great help would be a record of the temperature-dependence of the evolved compounds (i.e. the resolution of pyrograms with respect to temperature). It would simplify the pyrograms because different peaks appear at different temperatures. The first attempt to obtain that kind of resolution was made in stepwise PyGC, which is a multiple-input variant of conventional PyGC. Thermochromatography is a more advanced technique for obtaining temperature resolution. Both methods will be described in detail below.

5.1.2.2 *Multiple pyrolysis gas chromatography*

To record a representative evolution curve for each degradation product as a function of temperature, by using gas chromatography, the sampling theorem (Section 3.2.1) should be taken into account. When conventional packed and capillary columns are used, this is not an easy task, because of the long separation times. Multiple PyGC is an attempt to overcome the restrictions of long separation time by stopping the degradation reaction during the time that the separation is going on in the column. This is achieved by cooling down the sample during that period. The main assumption of the technique is that the polymer remains unchanged during the cool periods. From the theoretical point of view this is certainly not true, but in practice the sample does not change much if the impulses are not very intensive (i.e. not very long) and an approximate product evolution-rate curve can be recorded. The multiple-input technique in pyrolysis gas chromatography has arisen from the need to obtain more information about the gas evolution profile during the same run. There are some methods which are named differently but serve the same purpose. The stepwise pyrolysis technique was thoroughly described by Lehre and Robb in 1967 [20]. They studied the degradation mechanism and calculated the degradation kinetic parameters for poly(methyl methacrylate), polyacrylonitrile and polymethacrylonitrile. The analysis of evolved gases obtained by them is presented in Fig.

5.10. Each rectangle represents a separate chromatogram. The sample was mounted in a spiral filament heated first for 10 sec up to 150°C, and the chromatogram was recorded. This was repeated in 100°C steps and in each case, the effective sample was the residue remaining after the preceding heating. The pyrograms obtained in this way were named temperature-series programs. The pyrogram series were also used in the characterization of polymers and in kinetic calculations. Several modifications of multiple-input techniques were introduced at approximately the same time.

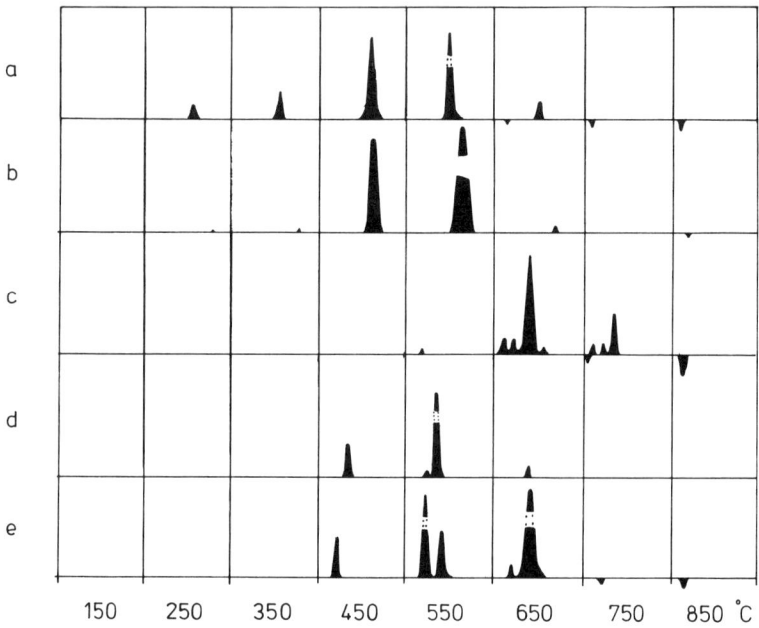

Fig. 5.10 — Temperature-series pyrograms. The same sample was pyrolysed 8 times at different temperatures. Polymers: a, poly(methyl methacrylate); b, poly(ethyl methacrylate); c, poly-(methyl acrylate); d, polystyrene; e, methyl methacrylate–styrene mixture (50/50). (Adapted from [20]).

Single-input conventional pyrolysis and stepwise pyrolysis gas chromatography were compared by Sugimura et al. [21] to provide structural information about ethylene–methyl methacrylate random copolymer. The composition of the polymers was also determined by elemental analysis and IR spectrometry. Stepwise pyrolysis was combined with the thermogravimetric method. The polymer sample was pyrolysed by use of an infrared image furnace coupled with a thermobalance. The sample was mounted on the sample holder and pyrolysed stepwise by applying pulsed heat radiation from the image furnace, by which the temperature of the sample was changed rapidly to any desired temperature up to 1000°C at a rate of up to 50°C/sec. During each shot of the pulsed radiation, the changes in the sample weight and temperature were recorded simultaneously and the evolved gases were transferred by the carrier gas into the gas chromatograph. Another version was to collect,

Sec. 5.1]   **Characterization of gases evolved from polymers**   159

after each shot, the pyrolysis products in a trap which was a capillary column (0.5 mm i.d., 20 cm long) inserted between the pyrolysis unit outlet and the separation column. The trap was cooled in ice-water during the pyrolysis and afterwards was heated to the column oven temperature and then programmed from 25 to 80°C at a rate of 20°C/min.

In selection of suitable conditions, stepwise pyrolysis provided a series of pyrograms for a given polymer sample. One analysis involved 20 heat pulses at 400, 420 and 440°C. The recorded results for the ethylene–methyl methacrylate copolymer consisted of the sequences of chromatograms, weight loss and pulse temperatures (Fig. 5.11). The column used was a 1.8 m copper tube (3 mm i.d.) packed with 30% Octoil-S oil on 80/100-mesh Celite 545. The initial oven temperature was 25°C, and was raised to 80°C and held there. During the preheating (to 80°C) four light C1–C4 components were separated. The cumulative weight loss showed that 10–12 heat pulses at 400°C were sufficient for most polymers. The pulse temperature remained constant during the cool periods. This kind of combined data presentation gives a good survey of polymer degradation but the data-handling and presentation are quite time-consuming.

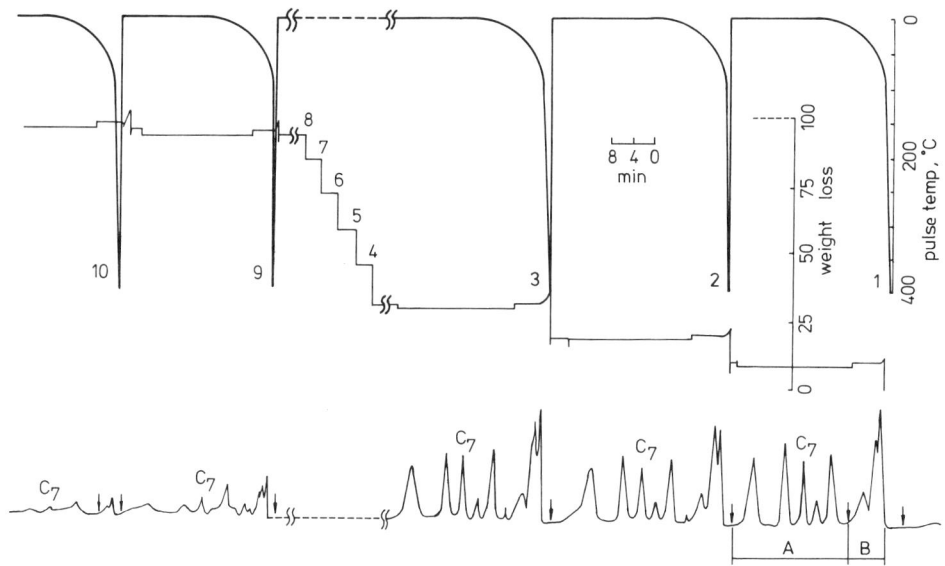

Fig. 5.11 — A series of pyrograms obtained by stepwise pyrolysis of a random ethylene–methyl methacrylate copolymer, with 20 repeated 400°C heat pulses. Analysis cycle A+B. A, isothermal at 80°C; B, 0–80°C. At the same time, weight loss and pulse temperature were determined. (Reprinted with permission from Y. Sugimura, S. Tsuge and T. Takeuchi, *Anal. Chem.*, 1978, **50**, 1172. Copyright 1978, American Chemical Society).

For the most part the same technique was used by Ericsson [22] who studied the decomposition kinetics of *cis*-1,4-polybutadiene. The technique was called sequential pyrolysis and the aim was to show the possibilities for pyrolysing a substance in a limited temperature interval to achieve temperature equilibrium conditions, which would be characteristic for the substances, depending on the degradation rate. The main differences between the Ericsson system and previous ones were in the pyrolyser, which was of the filament-type. The pyrolyser consisted of a 0.012 mm platinum foil (15×2.6 mm) heated by two current pulses. The first pulse lasted 8 msec for all pyrolysis temperatures and heated the foil to the desired pyrolysis temperature. The second heating pulse maintained the pyrolysis temperature and compensated for heat losses through the walls of the pyrolysis unit. The temperature was calibrated by means of a photodiode placed under the foil. At low temperatures the resistance of the foil was used to determine the temperature. The pyrolysis chamber was preheated to 115°C, which also compensated the heat shock. For sample preparation 14.7 mg of 96.5% cis-1,4-polybutadiene was dissolved in 2 ml of toluene. Then 1 μl of solution was placed on the heatable foil by means of a microsyringe. Before pyrolysis the solvent was evaporated by heating the foil.

The results obtained by sequential pyrolysis of a rubber sample are presented in Fig. 5.12. The sample was exposed to the same temperature at each pyrolysis in the series. After each heating the evolved gases were separated in the gas chromatographic column and recorded.

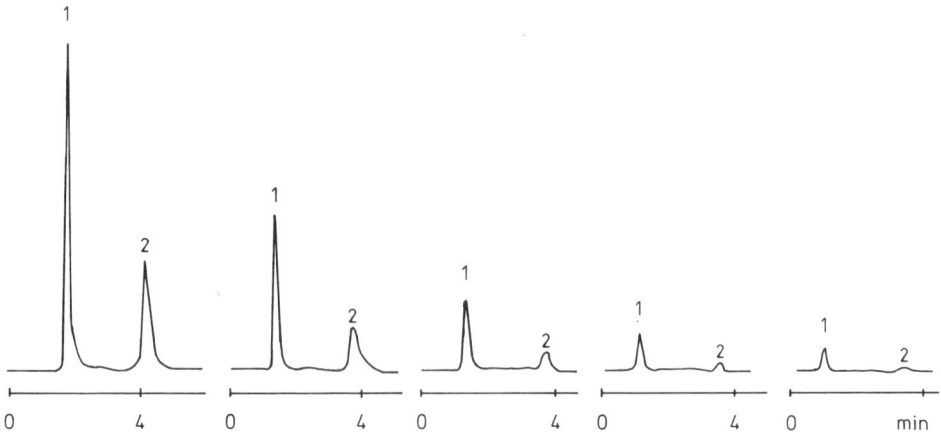

Fig. 5.12 — 7.4 μg of cis-1,4-polybutadiene pyrolysed sequentially at 450°C for 2 sec. Peaks: 1, butadiene; 2, vinylcyclohexene. (Adapted from [22]).

The initial sample was 7.4 μg and was pyrolysed at 450°C for 2 sec. The separation of the evolved gases after each pyrolysis cycle took 5 min. The total number of cycles is not given but the author notes that it depends on the time required to decompose all the sample, as indicated by the cessation of monomer production. Mainly, the total number of repeatable pyrolyses depends on the thermal stability of the substance and is specific for each particular substance.

The sequence of chromatograms allows us to observe the evolution of butadiene and vinylcyclohexene, the only two peaks which allow quantitative calculations. Here is evidence of the need to optimize the pyrolysis gas chromatographic analysis conditions. The column used was stainless steel, 3 m long, 1.9 mm i.d., stationary phase 15% Apiezon L on Chromosorb W, 80/100-mesh, AWE or OMCS, temperature 150°C. In this type of experiment the separation ability is inadequate. To overcome this problem, the amount of sample or the sensitivity of the gas chromatograph needs to be increased. Increasing the practicable amount of sample by using the Curie-point pyrolysis system is limited, and increasing the sensitivity gives rise to an increase in the signal noise, which affects the precision of the calculation.

One simplified approach to sequential pyrolysis is the so-called one-step two-shot analytical technique adaptation of pyrolysis gas chromatography [23]. To differentiate that method from conventional pyrolysis gas chromatography, the author named it chromatopyrography but this has met with some criticism of the terminology [24]. The main idea of this technique is that during the first thermal pulse the features of the volatile ingredients are characterized. The author introduced this procedure because it is sometimes more important to know the volatile ingredients than the polymer itself. On the other hand, as mentioned by the author, the volatile ingredients may be lost during the waiting period in conventional pyrolysis gas chromatography and the remaining volatiles will cause irreproducibility of the results. The second pulse identifies the polymer structure.

The whole analysis consists of the following operations. The injection port is preheated to 270°C and maintained at that temperature, and the analysis begins as soon as the sample is inserted. This first pulse drives out all the volatile ingredients and gives the first pyrogram. After the end of the first analysis and once the baseline has stabilized, the 'thermally purified' sample is pyrolysed at 1000°C for 15 sec to develop the second pyrogram, which characterizes the purified polymer. Finally, the polymer residue can be analysed by elemental analysis. The method would seem to be effective in giving a quick answer as to whether or not a polymer contains ingredients that are vaporizable at 270°C. The author confirms that all volatile ingredients are vaporized at 270°C but does not take into consideration the possibilities that some polymers may start to be degraded at below 270°C. The one-step two-shot analytical technique described may be classified as stepwise pyrolysis.

The three papers reviewed were the first to report the introduction of the multiple-input technique into pyrolysis gas chromatography. Compared with the conventional pyrolysis gas chromatographic technique, sequential pyrolysis and stepwise pyrolysis give more information about the thermostability of high molecular-weight compounds and enable kinetic calculations to be made. The one-step two-pulse technique is a double-input pyrolysis gas chromatography and can find a limited but successful use in process control or material identification and characterization in industrial quality control and test procedures.

In practice, the sequential and stepwise techniques do not differ from each other. Increasing the number of pyrolyses (the number of heat pulses) with visual interpretation of the results (chromatograms) and handling the data later turns out to be time-consuming and troublesome. All the above-mentioned studies were made before the 1980s and at a time when the increase in the use of artificial intelligence in analytical instrumentation and chromatography had not affected the quality and speed of analysis.

As seen from Section 4.2.2.2, the pyrolysis temperature is a critical and important parameter in pyrolysis gas chromatography. Figure 5.13 shows two profiles of the pyrolysis temperature in a sequential or stepwise case. A 'good' profile is one in

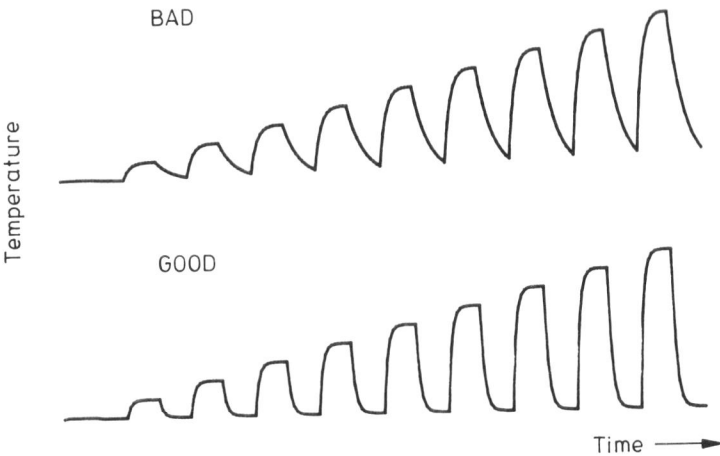

Fig. 5.13 — Temperature profiles in stepwise pyrolysis.

which pyrolysis temperature in the reactor rises abruptly to the maximum temperature required and after being held there it drops as quickly as possible to the initial level. A 'bad' temperature profile is one in which heating and cooling take some time and the reactor temperature never reaches the initial value after cooling. To overcome this problem, continuous linearly increasing sample heating is the best practical solution and will be considered in the following section on 'thermochromatography'.

5.1.3 Thermochromatography

Most studies (or papers) on pyrolysis gas chromatography assume that it is a powerful tool in analytical chemistry and particularly, in polymer characterization. They mostly deal with conventional pyrolysis gas chromatography in which the detector output gives time-resolved peaks which correspond to the substances separated from the mixture. The single-input technique gives a one-dimensional presentation of results in the form of a gas chromatogram. The modified technique is multiple-input (e.g. sequential) pyrolysis gas chromatography which has the following features: one sample is used, with multiple temperature pulses resulting in multiple analysis and a series of one-dimensional outputs in the form of a sequence of chromatograms. In principle, this is a two-dimensional method although it is never claimed as such. Thermochromatography is a multiple-input method with the exception that the temperature of the sample increases linearly in the reactor and, simultaneously, the multiple chromatograms of the products are taken at equal intervals of time. The output is two-dimensional, composed of time- and temperature-resolved gas chromatograms.

Temperature is the most important factor in thermal degradation reactions, and to get reasonable results in reaction rate measurements, careful control of the reactor temperature is needed. As seen from the last paragraph and Section 4.2.2.2, stepwise pyrolysis does not guarantee a well defined temperature of the sample and we cannot be sure what is the real pyrolysis temperature during repeated analyses. It is preferable to heat the sample by increasing its temperature linearly (as in thermogravimetry and in evolved gas analysis using mass or IR spectrometry), because, as demonstrated in Section 4.2.2.2, temperature control is then easier. By running chromatograms during the course of heating, we can obtain the reaction kinetics for each reaction product. This is a 'reverse' mthod of thermogravimetry which registers not the sample weight loss but changes in the amount of gas evolved. The second dimension of the method arises from the chromatographic separation of the evolved products. This gives the possibility of identifying the reaction products. To differentiate this evolved gases analysis system from conventional pyrolysis gas chromatography, we suggest using the term thermochromatography' (ThGC) because the output signal is a function of two independent factors: the reactor temperature and the chromatogram running time.

Thermochromatography can be considered a modified pyrolysis gas chromatographic method where the main aim is to overcome or smooth the shortcomings which appear in conventional pyrolysis gas chromatography, by applying, to a considerable extent, system control, data recording and handling, and graphical presentation of results, by on-line computer. The backbone of thermochromatography is the sample multiple-input principle that assumes reorganization in the normal gas chromatograph inlet system and data presentation. Because of this, a computer is an essential part in the thermographic analysis system.

5.1.3.1 *Scope of thermochromatography*
Depending on the nature of the GC detector signal, thermochromatography can conveniently be used in two different modes.

(1) Correlation gas chromatography may be used if the detector noise level is comparable with the output signal. The data handling and representation require the use of a computer and special mathematical methods. This is the case when the rate of evolution of the reaction products is low (i.e. in the 'low-temperature' region). The number of components evolved is small, resulting in simple chromatograms. It is difficult to fulfil the sampling theorem requirements because long pseudo random binary sequences (PRBS) are necessary to obtain a good signal-to-noise ratio and, as already known from Chapter 2, the PRBS duration should be equal to the chromatogram length. On the other hand, if the degradation rate is not high (low heating rate), direct measurements are possible and the amount of products obtained can be averaged over the PRBS duration.
(2) Equi-interval sampling with high-speed chromatography is used if the sample is significantly degraded and many products are evolved at a high rate. Noise is then no longer a problem, but the high evolution rate of the products and the large variations in their amount requires a rapid chromatographic analysis. The temperature interval in which this occurs we call the 'medium-temperature' region. The role of ThGC in evolved gases studies is given in Fig. 5.14.

The choice between using correlation and high-speed chromatography is made on the basis of the signal-to-noise ratio. The boundary between the two is indistinct

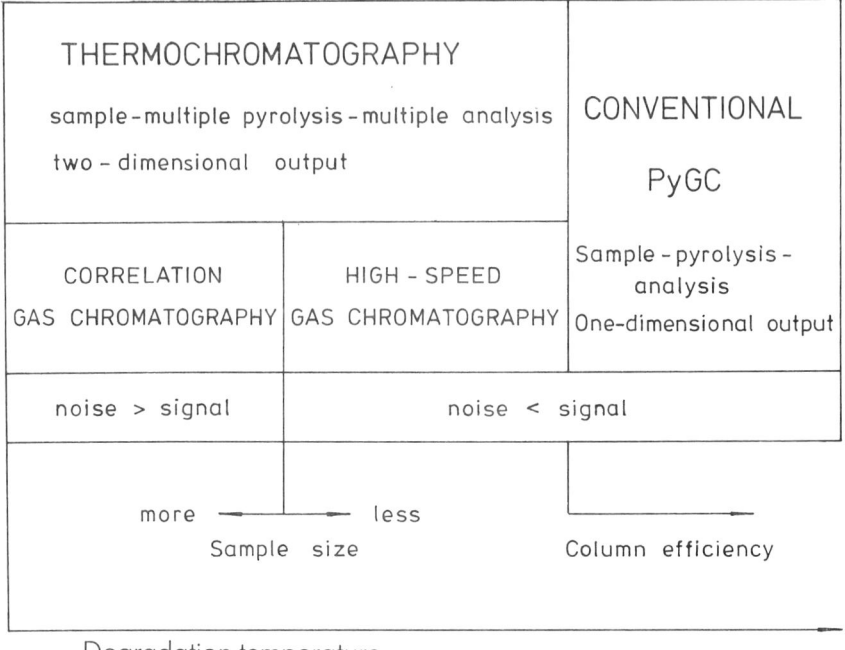

Fig. 5.14 — Conventional pyrolysis gas chromatography and thermochromatography as a function of sample size, degradation temperature and column efficiency.

and depends on the nature and amount of the sample (which rarely exceeds 100 mg). The choice is in effect an exercise in optimizing the analysis conditions, and it depends on the thermal stability of the sample. In practice, the extent of degradation of a particular sample is the limiting factor that determines the input sequence used. Therefore, we must pay attention to two cases. (a) The sample is degraded easily, giving a large number of decomposition products which are readily detected by the gas chromatographic detector. The correlation chromatographic region then shifts to that of lower temperatures. (b) Samples which are relatively thermally stable, giving small amounts of degradation products, shift the application of correlation chromatography to the region of higher temperatures. Despite the difference between them, the correlation and high-speed gas chromatographic techniques use the same analytical equipment.

The upper boundary of high-speed chromatography application is also indistinct and depends on the separation ability of the column. If the sample begins to degrade at a high rate, giving an increasing number of components, it is difficult to separate them within the interval of time required by the sampling theorem, and generally only a small number of chromatograms can be recorded during the heating of the sample in the high-temperature region. In reality this is the conventional pyrolysis gas chromatography area.

The distribution of the degradation temperature ranges is somewhat arbitrary, but it may be classified as follows: 'low', from room temperature to 200°C; 'medium',

200–550°C; 'high',>550°C. In general, the separation ability of the column and the character of the evolved gases are the most decisive factors in choosing the analysis parameters.

In comparison with mass spectrometry and IR spectrometry, thermochromatography (also pyrolysis gas chromatography) is the only method that gives a real picture of the changes in the amounts of evolved gases. In mass and IR spectrometry the final results are calculated on the basis of compound fragments or spectrum absorption bands.

5.1.3.2 *Experimental conditions in thermochromatography*

In conventional pyrolysis gas chromatography, all the gases evolved during degradation must be directed to the analytical column at once. In thermochromatography, the time between injections of evolved gases differs in correlation and high-speed chromatography (Fig. 5.15). The injection time intervals increase in the sequence:

$$\Delta t_{correlation} < \Delta t_{high-speed} < \Delta t_{conventional}$$

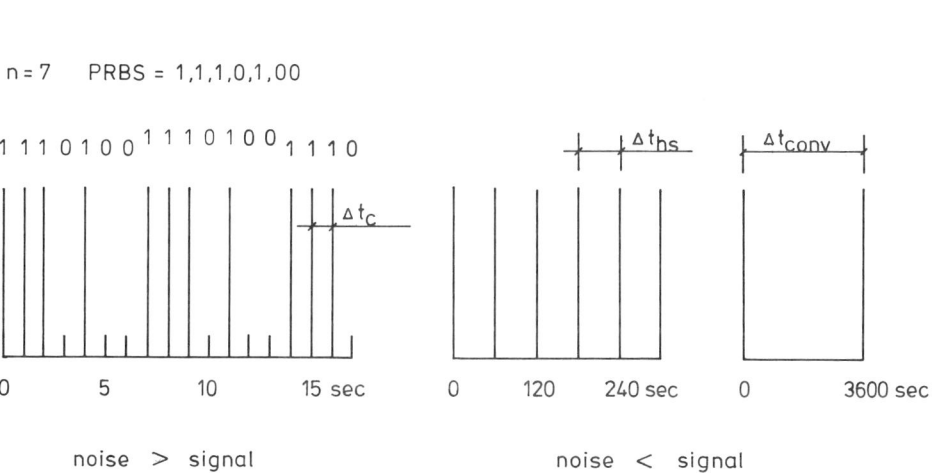

Fig. 5.15 — Sample injection interval in correlation, high-speed and conventional gas chromatography. PRBS — pseudo random binary sequence.

Δt_{cc} for correlation chromatography $< \Delta t_{hs}$ for high-speed GC $< \Delta t_{con}$ for conventional PyGC. The mean numerical values of the time intervals between injections for this sequence are $\Delta t_{cc}=0.1$–1 sec, $\Delta t_{hs}=20$–60 sec, and $\Delta t_{con}=1$–60 min. The analysis time for the conventional system is based on the assumption that the pyrolysis temperature is $>500°C$ and a great number of peaks are formed; such a system commonly needs about 60 min per chromatogram (single input).

The injection volume is determined by the injector valve construction, the switching time and the carrier gas flow-rate. In thermochromatography, the evolved gases are taken for gas chromatographic analysis (ThGC) according to a previously determined rule that is different for the correlation and high-speed modes. In the high-speed method the time intervals between injections are constant, usually 1–2 sec but even 0.1 sec is possible if needed.

In the correlation method the gas flow scheme is the same as for the high-speed mode, but the injections are regulated by a pseudo random binary sequence (PRBS). The injection interval may be 0.5–5 sec. As already known from Chapter 1, PRBS is a sequence of zeros (no injection) and ones (injection) distributed in a seemingly random sequence. In fact, PRBS is not a random sequence, but is determined and generated according to certain rules.

During injection the collected evolved gases in ThGC are directed to the gas chromatographic column for separation. At the same time, the degradation of the sample in the reactor continues and the products are vented from the pyrolyser to the air. To plan a study of polymer degradation as a function of temperature, it is useful to foresee the following: (1) the range of degradation temperatures; (2) the number of points for describing the chromatogram; (3) the sampling interval.

The data-recording and presentation capabilities depend on the computer memory. The capacity of the computer memory limits the data acquisition, and attention must be paid to economical use of computer resources. If the data are recorded in two bytes, the chromatogram is recorded for 256 points and 40 chromatograms are required, then the necessary computer memory is $2\times256\times40=20$ kbytes. If the memory resources must be shared with the operating system and graphics pages then the capacity needed would exceed the memory of a 64 kbyte micro. However, use of any kind of memory buffer or of additional cards would help.

The loss of information can be high if the width of the sampling temperature interval has been improperly chosen. The sampling temperature interval is determined here as the sampling time interval multiplied by the sample heating rate. The effect of an improper sampling temperature interval can be demonstrated by the following example, based on the conditions for a thermochromatographic analysis of an isoprene rubber and of a mixture of nitrile rubber (NBR)+chloroprene (CR), given in Table 5.2.

Table 5.2 — Analysis conditions for rubber ThGC

Analysis conditions	Sample		
	Isoprene	NBR+CR (A)	NBR+CR (B)
Degradation range, °C	250–500	250–500	200–270
Heating rate, °C/min	20	20	1
Analysis time per chromatogram, sec	58	58	58
Sample injection time, sec	0.5	0.4	0.4
Number of injections in sequence	37	37	37
Number of points on chromatogram	256	256	256

The stack plot of thermochromatograms of isoprene rubber degradation, shown in Fig. 5.16C, gives quick visual information about the thermal stability of the

Sec. 5.1]   **Characterization of gases evolved from polymers**   167

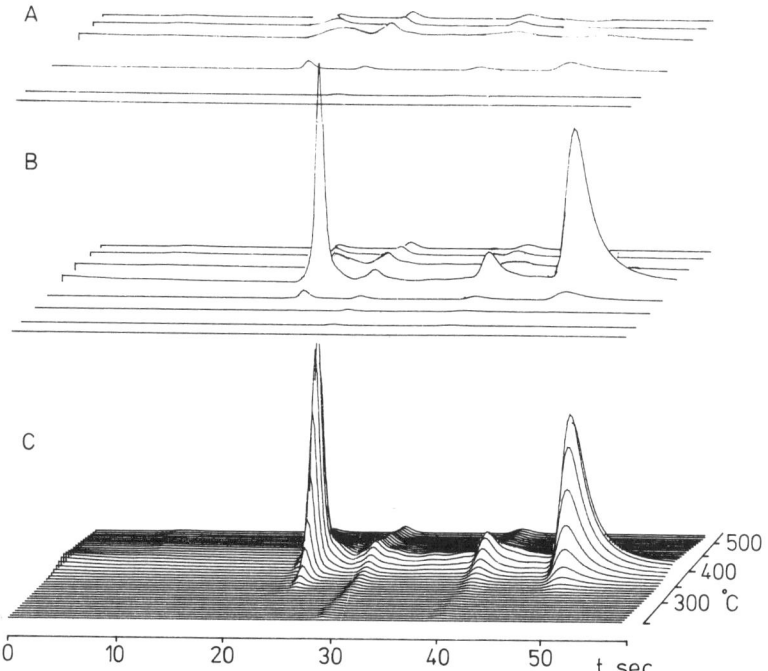

Fig. 5.16 — Typical thermochromatograms of isoprene rubber degradation in the temperature interval 250–500°C. Too big an interval of pyrolysis temperature may lead to a false conclusion about the thermal stability of the sample. Only small temperature differences ($\Delta T=20°C$) allow satisfactory description of sample degradation (C). Other temperature intervals: (B), 50°C; (A), 100°C.

isoprene rubber and shows that a temperature interval of 20°C or less is needed between samplings. If the pyrolysis temperature interval is 100°C, Fig. 5.16A, we can reach a wrong decision that the isoprene rubber is a good thermally stable material which is not degraded much in the temperature region 250–500°C. Taking the sample at intervals of 50°C (Fig. 5.16B), it appears that the thermal stability of isoprene rubber is worse than it seemed at first glance. Exhaustive information about thermal degradation can be obtained if the sampling interval is 1 min, corresponding to a temperature interval of 20°C. The same situation is shown in the sequence of thermochromatogram slices (Fig. 5.17). For kinetic calculations the evolution curve shape and the total amount of each individual evolution product can show the real situation and therefore can be determined at the maximum frequency allowed by the computer memory.

If it is necessary to study the degradation process in more detail then a lower temperature-rise speed is needed. The effect of heating rate on the sampling temperature interval is demonstrated in Fig. 5.18. The degradation rate of the acrylonitrile–butadiene copolymer+chloroprene rubber (NBR+CR) increases abruptly in the temperature interval 220–230°C and the heating rate of 20°C/min is too high to give a good image of what really happens in this region (Fig. 5.18A).

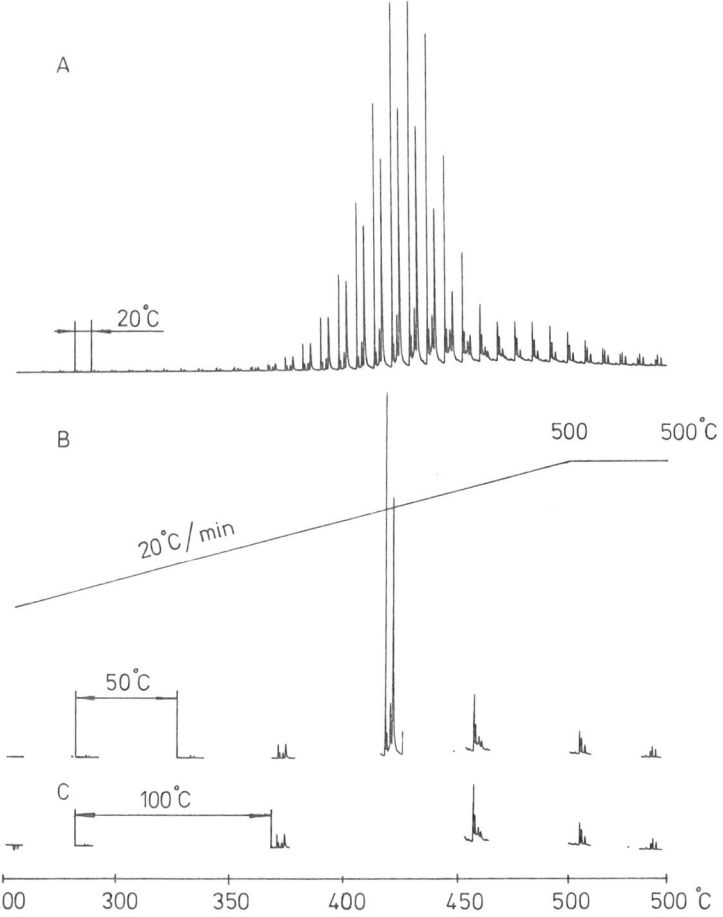

Fig. 5.17 — Loss of information. Results are presented as thermochromatogram slices; initial data are the same as for Fig. 5.16.

However, lowering the heating rate to 1°C/min allows the necessary number of measurements (Fig. 5.18B).

The main task in ThGC is to obtain maximum separation of components in minimum time. From that point of view the advantage of the open tubular columns (PLOT, SCOT and WCOT) is that separations can mostly be performed at a higher speed and lower temperature than with packed columns. Also, the precision of retention data is better with open-tubular columns than with packed columns. The ability of the Deans-type flow switch (Section 4.3.2) to perform gas sample injection with a precision of 0.3% and a duration as short as 0.1 sec, gives good possibilities for using open tubular columns in thermochromatography.

The column temperature is an important variable. The column temperature selected must be high enough to enable elution of the evolved gases in a reasonable

Sec. 5.1]   Characterization of gases evolved from polymers   169

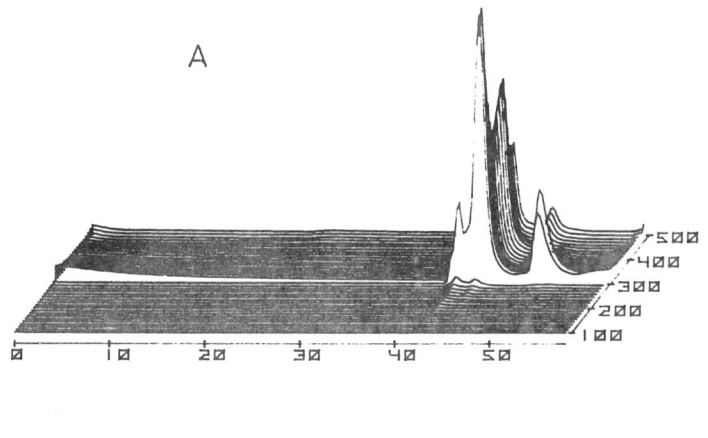

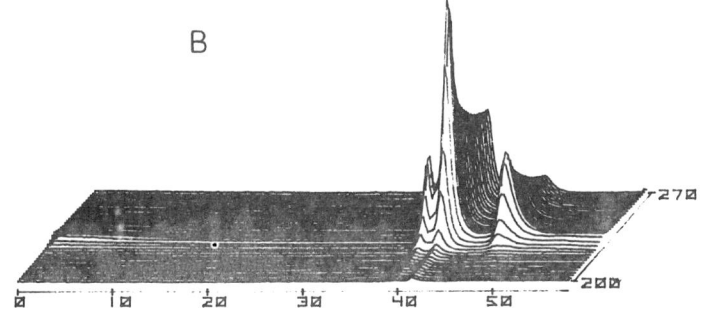

Fig. 5.18 — Loss of information if the temperature interval is too wide. A, Too high a temperature rise (20°C/min); B, temperature rise 1°C/min. Anaylses were performed on different samples.

time but not so high that column bleed or incomplete separation becomes a problem. In ThGC, the column oven normally operates in the isothermal mode because temperature programming contains several cycles needing manipulation of temperature during rather a short cycle. In high-speed thermochromatography the cycle time and oven temperature fluctuation may cause a cumulative shift in chromatogram sequence, that can give fluctuating results.

The ThGC apparatus system (Fig. 5.19) is based on multiple-input chromatography concepts. The sample is placed in a small silica tube which is inserted into the reactor. The inert gas, which is also the carrier gas, normally flows through the reactor and carries the evolved gaseous products through the sampling value to the atmosphere. Pure carrier gas also flows through the other path in the sampling valve, into the gas chromatograph. At a command from the computer, the Deans-type valve reverses for 0.1–5 sec. and the gas stream, which contains the evolved gases, is directed into the gas chromatographic column. The detector output analogue signal is converted into a digital one and stored on a floppy disc for future use. It is helpful to record the detector analogue output on a chart recorder as well. This primary record is useful for choosing proper analysis conditions: the range and rate of change of the

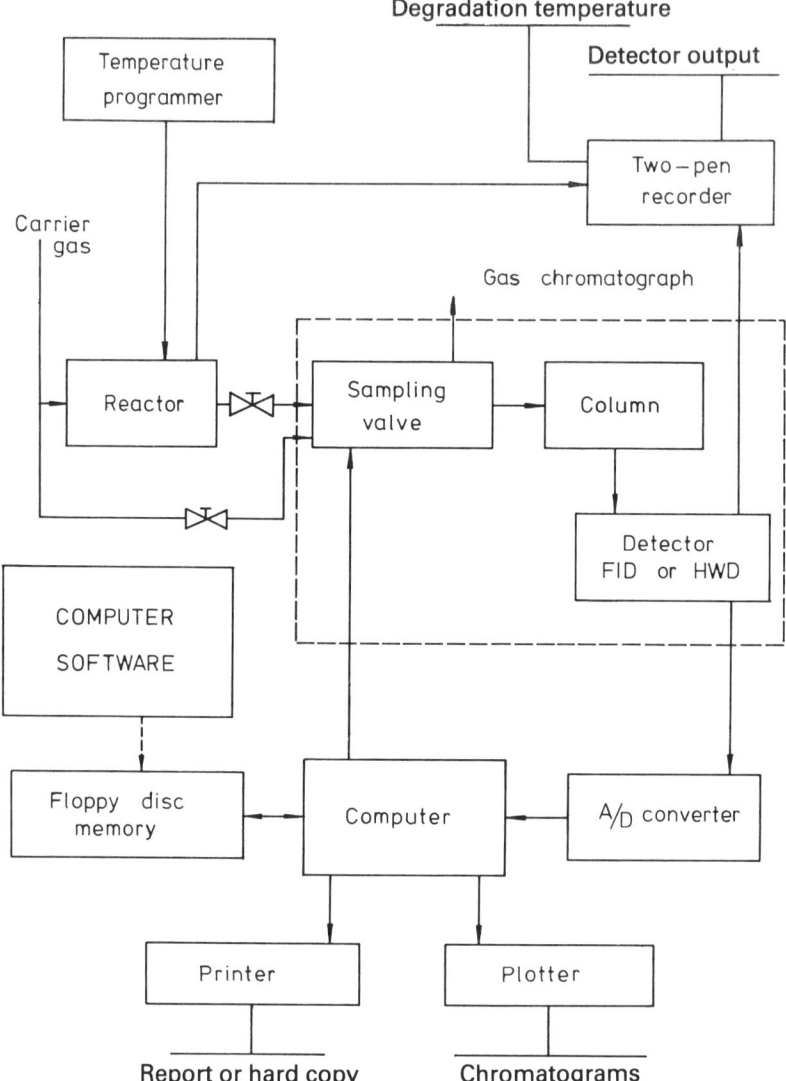

Fig. 5.19 — Block diagram of computer-based multiple-input gas chromatography. The system can be used in both thermochromatography (correlation and high-speed chromatography) and conventional pyrolysis gas chromatography.

degradation temperature, sample size, sample injection time, and time interval between injections. If a two-channel recorder is available, the second channel can be used to record the reactor temperature.

The reactor temperature is programmable from 50 to 700°C, at heating rates ranging from 1 to 20°C/min. The reactor temperature is controlled by a special temperature controller or computer. As the amount of gases evolved depends on the

Sec. 5.1] **Characterization of gases evolved from polymers** 171

sample weight, and is specific for each polymer, the active volume of the sample container in the reactor must allow analysis of 1–100 mg samples.

Although many new powerful personal computers are commercially available today, not all of them are well suited for controlling small experimental arrangements. The Apple IIe personal computer is quite suitable for the kind of system-control and data-handling described above. One of its important advantages is that the Apple IIe has several slots in which application-dependent interface cards can be inserted. It is also relatively easy to write a program to control the experiment by using the 6520 P (microprocessor) code and Basic (Applesoft) language.

The computer software for thermochromatography can be divided into two parts: (1) the system-control package which contains a program for sample switching and degradation temperature control; (2) the data-acquisition programs which read the analogue data, convert analogue into digital, store data in memory and on a floppy disc, transform digital code to an integer format, and plot chromatograms. There are three special programs for plotting which give: (1) a sequence of thermochromatogram slices, (2) a stack plot of thermochromatograms, and (3) contour plots of thermochromatograms.

5.1.3.3 Earlier applications of thermochromatography

Liebman *et al.* [25] were probably the first to utilize ThGC principles to characterize polymer degradation products. They reported the application of 'time-resolved' pyrolysis gas chromatography combined with a derivative thermogravimetric analysis, to a series of poly(vinyl chloride) homopolymers with different degrees of branching and also to a model copolymer series with a low amount of propylene in the chloride chain. In the time-resolved pyrolysis gas chromatograph, a sample of 2–4 mg was placed in a silica insert in the Pyroprobe-type pyrolyser and was heated at a selected rate (5, 10, 20 or 40°C/min) in an air or nitrogen atmosphere, or pulsed to any selected temperature in a rise-time of milliseconds. The pyrolysis system enabled a temperature of 1000°C to be reached. The degradation product flow-path was controlled by a six-port sampling valve which allowed automatic repetitive sampling every 2 min. The system used was a combined two-stage system in which pyrolysis and analysis were separated from each other, permitting pyrolysis not only in the carrier gas but also in different atmospheres. The pyrolysis products were separated during a 2 min cycle in a 12-ft 1/8-in. column, packed with UCW-98 (10% by weight on Chromosorb W) which separated benzene from toluene for later calculations of their ratio.

The selectivity of time-resolved pyrolysis gas chromatography can be seen from Fig. 5.20 where differences in polymer structure become evident. Polymers A and B are similar in overall composition. The difference between the two systems is as follows: in A, methyl units are randomly dispersed along the vinyl chloride chain and give a distinguishing feature not seen in sample B; B, according to ^{13}C NMR data, is a polymer containing a low amount of chloromethylene branch structure along the homopolymer chain. The benzene and toluene peaks were observed to differentiate these two systems. Time-resolved pyrolysis gas chromatography gives a survey of the benzene-toluene formation up to 600°C, with additional isothermal heating for 7 min at 600°C. The heating rate was 20°C/min, under nitrogen. The resolution of benzene and toluene is not as good as it might be, and there is a need for a more detailed study

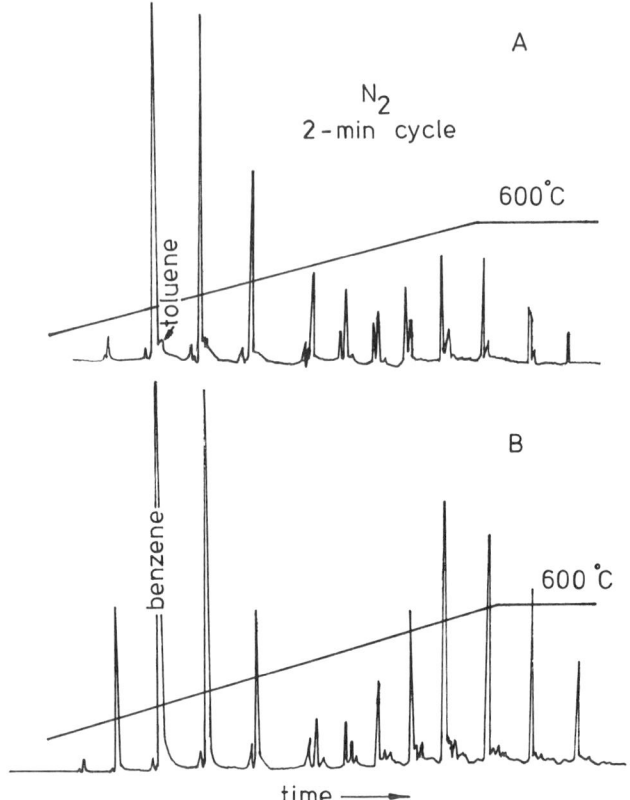

Fig. 5.20 — Results of an early application of time-resolved pyrolysis gas chromatography. Time-resolved peaks for two polymers (A and B) which are similar in overall amount and rate of product evolution. Heating rate 20°C/min. (According to S. A. Liebman, D. H. Ahlstrom and C. R. Foltz, *J. Polymer Sci.*, 1978, **16**, 3139. Reproduced by permission. Copyright 1978, John Wiley & Sons Inc.).

to be sure that the peaks contain only benzene and toluene. Resolution was decreased when the sample size was increased to 5 mg to register a detectable benzene peak at 250°C. This is typical of this kind of experiment. Increasing the sample size increases the signal for a low evolved gas concentration or in the low temperature region but causes poorer separation in the high-temperature areas where a large amount of thermal degradation gives a larger amount of evolved gas. To solve problems of this kind, system development and optimization of the analysis parameters are needed.

When the evolution rate of the degradation products is high, the sampling theorem requirement should be taken into account. In [26], samples of the evolved gases were collected in cold-traps. The minimum sampling time interval was governed by the time necessary to change the traps. Later the traps were connected to a chromatographic injection port and heated rapidly, and the evaporated compounds gave a chromatogram of the products obtained at the temperature at which the sample was degraded. This system was used in a study of the thermal degradation of polystyrene. The reactor itself was a DTA apparatus. Evidently, using traps in this 'off-line' ThGC system is an inconvenient procedure. The necessity to use tens of traps to record a reasonable evolution rate curve makes this method impractical.

These early applications demonstrate well the efforts of investigators to satisfy, by any possible means, the sampling theorem requirements in evolved gas analysis studies.

5.1.3.4 Rubbers

Rubbers are polydienes, an extremely important group of polymers. To this group belong natural rubber and its derivatives, and the products of polymerization and copolymerization of conjugated dienes.

The products of destructive distillation of rubber were studied by several investigators in the last century. Isoprene and dipentene were isolated many times, identified and shown to be the predominating products. Williams [27] and Bouchardat [28] were the first to study the thermal degradation of natural rubber. Later Midgley [29] destructively distilled, at atmospheric pressure, 90 kg of light pale crepe, in 7-kg batches. The temperature was raised as quickly as possible to 700°C. After the removal of isoprene, the residual oil was subjected to thorough fractional distillation, during which cuts were collected at every degree between 50 and 176°C. It was remarkable and tremendous work because even nowadays in pyrolysis gas chromatographic studies the degradation products are, in most cases, analysed only at intervals of 20, 50 or 100°C. Twenty-three compounds with 1–5 carbon atoms were identified. These experimental results together with earlier data were expected to throw light on the formula of the rubber molecule.

Nowadays, both natural and synthetic rubber are widely used polymers. A great variety of structural modifications afford them specific mechanical and chemical properties. A rubber compound is composed of one or more elastomers and ingredients. For example, the material used in the tyre industry is usually a complicated blend of rubber polymers. In addition, each rubber compound may contain about 10–20 different ingredients, such as sulphur, zinc, oxides, carbon black and antioxidants.

Two of the most important parameters of a rubber are its short and long term upper and lower (below 0°C) working temperatures (Table 5.3). The other characteristic parameters for rubbers are compressibility, elasticity, adhesiveness and elongation at breakage. Resistance to weather, ozone, acids, bases, aliphatic and aromatic oils, wear, and flame are also important. All these qualities depend on the polymer structure, as do the products of thermal degradation.

As can be seen, only two types of rubbers (fluorinated and silicone rubber) are capable of short term working at 275°C and in normal conditions at 200°C. For all other rubbers, the normal upper working temperature limit is 60–90°C which means that in most cases, long term working temperatures are well below 100°C. Owing to these properties and their wide application, rubbers are a good subject for thermochromatography. The overwhelming majority of rubber degradation studies have been made at temperatures above 500°C by conventional pyrolysis gas chromatography.

Table 5.3 — Rubber types and properties

Type	Working temperature, °C		
	Long term	Short term	Below 0°C
Natural	60	110	−30 −60
Styrene–butadiene	70	100	−20 −50
Butyl	90	150	−10 −40
Nitrile	80	130	−10 −50
Chloroprene	80	130	−20 −50
Urethane	60	80	0 −40
Fluorinated	200	275	−20 −40
Silicone	200	275	−50 −80
Chlorosulphonated polyethylene	90	150	−20 −40
Ethylene–propylene	90	150	−30 −60
Epichlorohydrine	90	150	−10 −50

Thermochromatography, as we saw in Section 5.1.3, can measure the degradation kinetics in the lower temperature region.

There are various methods for studying thermal degradation of rubbers: (1) direct degradation without previous purification of the sample; (2) rapid and complete vaporization of volatiles, fillers, etc. at a temperature of 270°C as the first step and then the final degradation of the 'purified polymer' at 700°C, as suggested in [23]; (3) extraction of soluble fillers, etc. by a solvent (usually methanol) in an extraction apparatus for 8–12 hr [30] and the complete removal of the solvent by heating in a vacuum oven at 60°C for 30 min. The last procedure is time-consuming and therefore a two-step pyrolysis seems to be a better method for obtaining pyrolysis gas chromatograms for the polymer freed from filler, etc.

Figure 5.21 is a pyrolysis gas chromatogram of a carefully purified rubber blend

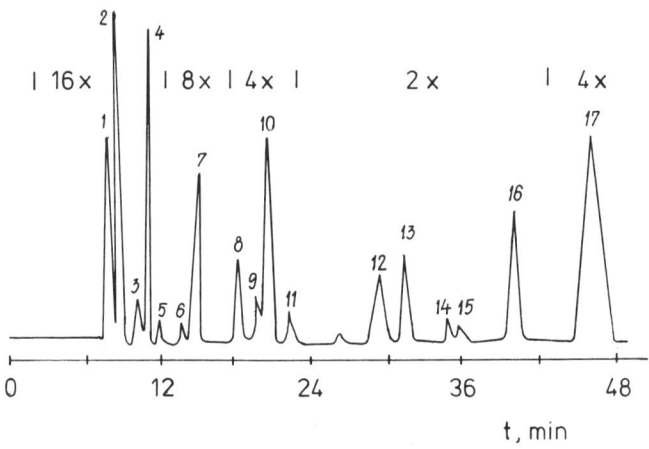

Fig. 5.21 — A pyrogram of a mixture of 40% natural rubber, 30% styrene–butadiene rubber, and 30% ethylene-propylene terpolymer rubber. (Reprinted with permission from A. Krishen, *Anal. Chem.*, 1972, **44**, 494. Copyright 1972, American Chemical Society).

which contained 40% natural rubber, 30% styrene–butadiene and 30% ethylenepropylene terpolymer rubber [30]. The blend contains three different types of rubber and it can be assumed that similar degradation products will be obtained from other rubbers. Separation was effected with an aluminium column (40 ft×3/16 in. o.d.) packed with 10% tricresyl phosphate on 60–80 mesh Chromosorb P. The pyrolysis temperature was 700°C and during a separation time of 54 min 17 peaks were separated and later identified. Peak identification and relative retention data are given in Table 5.4.

Table 5.4 — Peak identification and relative retention data on rubber degradation products. (Reproduced by permission, from A. Krishen, *Anal. Chem.*, 1972, **44**, 494. Copyright 1972, American Chemical Society)

Peak No. in Fig. 5.21	Compound	Relative retention (nonane=1.000)
1	Methane	0.0009
2	Ethane+ethene	0.0082
3	Propane	0.0268
4	Propene	0.347
5	2-Methylpropane	0.849
6	Propadiene+butane	0.0948
7	1-Butene+2-methylpropene	0.1048
8	*trans*-2-Butene	0.1409
9	*cis*-2-Butene	0.1649
10	1,3-Butadiene	0.1769
11	3-Methyl-1-butene	0.2074
12	1-Pentene	0.3038
13	2-Methyl-1-butene	0.3367
14	*trans*-2-Pentene	0.3848
15	*cis*-2-Pentene	0.3993
16	2-Methyl-2-butene	0.4476
17	Isoprene	0.5397

From the technological and engineering point of view, the thermal degradation of a rubber in the temperature region below 100°C is extremely important because it characterizes the changes in the rubber structure under working conditions. About 60% of all rubber produced is used in the manufacture of tyres, the working temperature region of which can be in the range 60–90°C. In that temperature interval degradation products are difficult to detect without preconcentration or collection and there are no studies in that region that are known to us. Studies of rubber degradation in the region 200–500°C indicate that at above 300°C the degradation increases enormously, producing characteristic thermochromatograms. Complete volatilization might occur in 30 min at 400°C [31] or immediately at above 500–550°C. Although thermal degradation in the input gas atmosphere of a gas chromatograph differs from the real working conditions for rubber in air, it gives some knowledge of the degradation process. Quite frequently, the aim of the studies is to compare similar blends, for which conventional pyrolysis gas chromatograms could have differences that are too small to be noticed. Another large area that has need of that kind of analysis is the monitoring of technological processes. There are good prospects for thermochromatography in solving that kind of problem, which essentially involves the separation and determination of evolved gas components.

The sphere of specific investigations in rubber chemistry and technology can be very wide. For example, aging in different climatic conditions, by heat and solar radiation, and exposure to chemicals, aerial oxidation and extreme conditions.

Figure 5.22 presents thermochromatograms which illustrates the possibilities

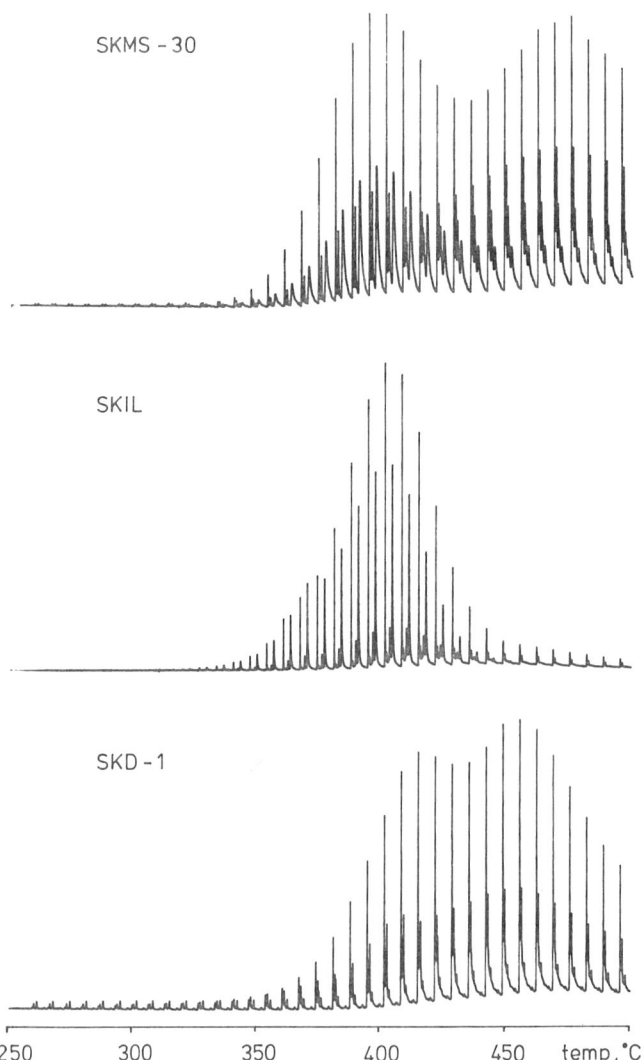

Fig. 5.22 — Sequences and thermochromatogram slices of several rubbers. Butadiene–methylstyrene copolymer (SKMS-30); isoprene rubber (SKIL); stereoregular 1,4-polybutadiene rubber (SKD-1).

Sec. 5.1]   **Characterization of gases evolved from polymers**   177

of the multi-input gas chromatographic technique in high-speed thermochromatography.

The individual peaks on the thermochromatograms have not been identified, because the main aim was to compare the thermal stability of different rubber blends and to show the applicability of thermochromatograms to qualitative analysis. The figure shows a sequence of thermochromatogram slices for three different types of rubber. Each sample was heated with a linear temperature increase from 250 to 500°C at 10°C/min. Each sequence contains 37 chromatograms. The analysis conditions were as follows: detector FID, column temperature 150°C, 0.5 mm i.d. 45 m stainless-steel wall-coated open tubular SE-30 capillary column, sample weight 1–2.5 mg. The sequence of thermochromatogram slices is the primary output recorded on a flat bed recorder. Besides the direct output, all data were stored in computer memory and later used in a different presentation. The rubbers under study were: butadiene–methylstyrene copolymer (SKMS-30) containing 22–24% methylstyrene, which begins to degrade at 250°C but undergoes rapid degradation at 400°C.

The isoprene rubber (SKIL) is more thermally stable than the SKMS-30, which is synthesized by using a lithium catalyst. Its structure is less regular than that of natural rubbers and it contains 5–6% of 3,4-polybutadiene structure. Below 350°C there is no sign of degradation. The stereoregular 1,4-polybutadiene (SKD-1) starts to degrade at 200–250°C, but intensive degradation takes place at 350–500°C.

Stack plots give easily understandable visual information about the degradation. Despite the fact that the sequence of thermochromatogram slices gives the same information, the stack plots are easier to understand. Figure 5.23 presents stack plots of the rubbers, with the chromatograms arranged in order of temperature rise. The SKMS-30 stack plot shows the cluster of peaks with retention times of 20–30 sec. (The peaks labelled S2 at the beginning of the plot are the tails of the peaks labelled S1, and are shown in this way for convenience.)

In the presentation of the stack plot it is possible for big peaks to hide small peaks which appear at a higher temperature. To avoid the problem, a presentation of the rear view of the same stack plot is needed. Although such peaks could be found from a sequence of thermochromatogram slices, the rear-view stack plot is easier to interpret. The rear-view stack plots of the thermochromatograms in Fig. 5.23 are given in Fig. 5.24. In the rear-view plot for SKIL the hidden peak (b) of Fig. 5.23 is apparent.

The third possibility for presenting data in thermochromatography is a contour plot, which is also very informative. The contour plots of the same three rubbers are presented in Fig. 5.25. The numbers beside the contours indicate the cut height. The value of the highest peak is taken as 1.0. The contour plots are very characteristic but their interpretation needs some experience. Another way of assessing contour plots is given in Fig. 5.26. The contour plots permit comparison of two polymers in the quality control of a production process. This figure presents two contour plots and the pattern obtained by subtracting one from the other.

Figure 5.27 gives thermochromatograms of pure butadiene–acrylonitrile copolymer (NBR) and chloroprene (CR) rubbers and their mixtures. The results allow us to make decisions about the thermal stability of pure and mixed rubber blends. Rubber blends are used as insulating material in the cable industry. The quality and durability of these materials depend to a great extent on their composition, and

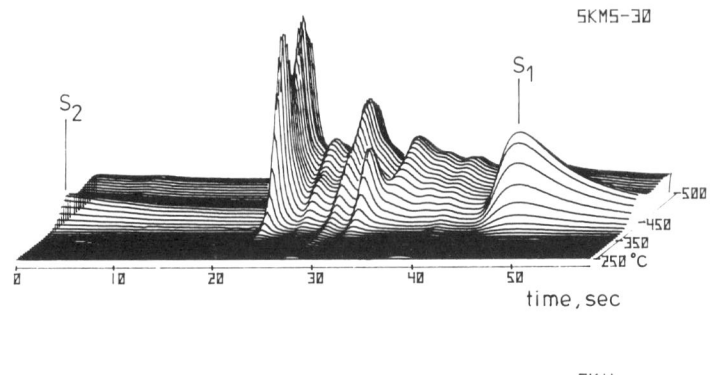

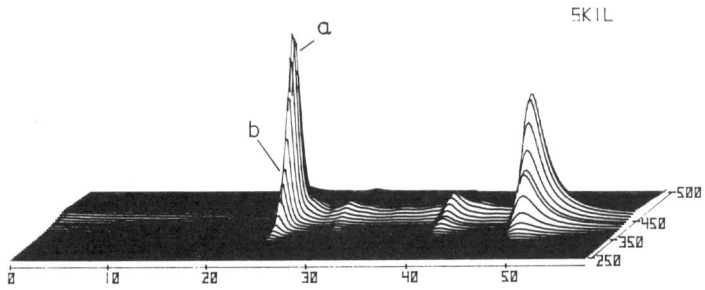

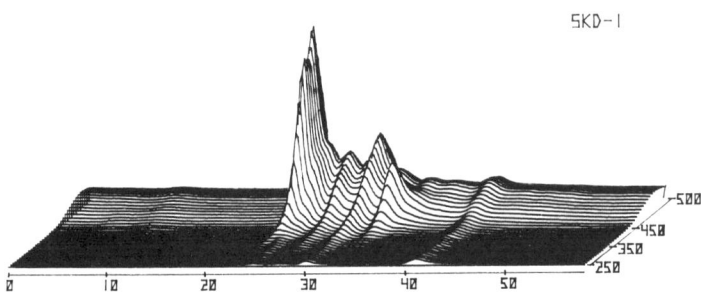

Fig. 5.23 — Stack plot of thermochromatograms. Another presentation from the same initial data as for Fig. 5.22. The thermochromatogram of SKMS-30 is 'folded back' for convenience (the cluster of peaks S2 is a continuation of S1; 'b' marks the location of a hidden peak shadowed by peak 'a' (see the next figure).

therefore their study is extremely important. The evolution of degradation products of some insulation materials is shown in Fig. 5.28.

5.1.3.5 High-temperature resistant polymers: the role of flame retardants

Bromine, chlorine, phosphorus and boron compounds and some other chemicals, such as antimony oxide and alumina trihydrate etc., are known as flame retardants. The vapour-phase reactions of the bromine and chlorine compounds inhibit the chemical reactions of combustion. Alumina trihydrate has several features, such as

Sec. 5.1] **Characterization of gases evolved from polymers** 179

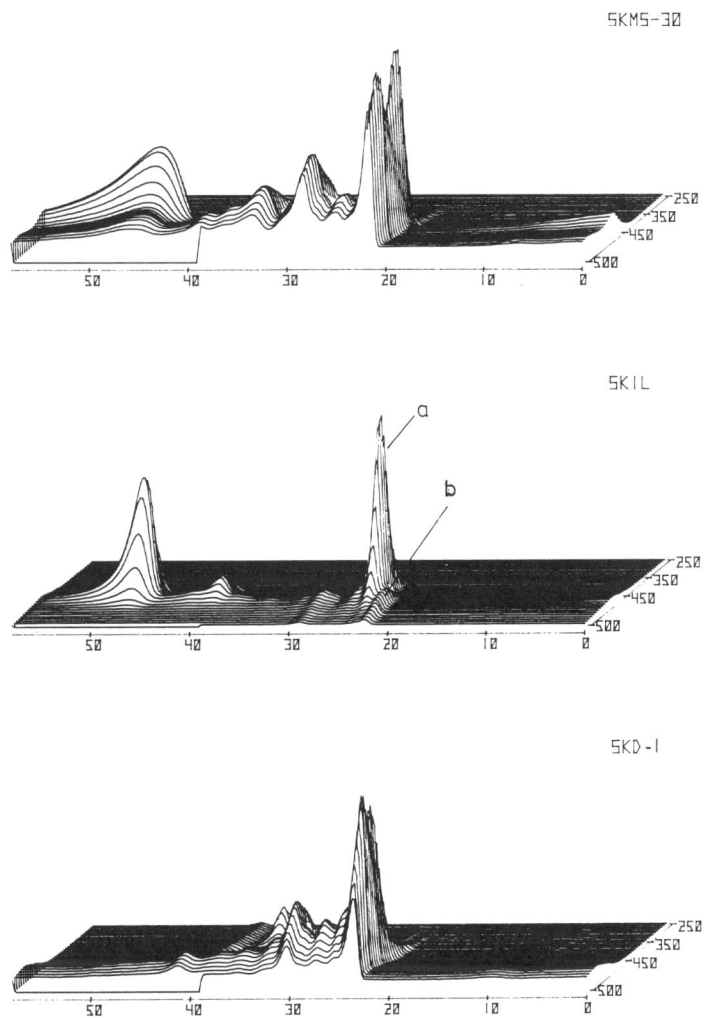

Fig. 5.24 — Rear-view stack plot of the thermochromatograms presented in Fig. 5.24. In this view, the hidden peak 'b' is totally in sight.

low smoke production and non-toxicity. Its flame-retardant mechanism is endothermic (heat absorption), the flame being cooled by the process of dehydration of the hydrated alumina.

Antimony oxide is largely used in polymers when halogen compounds are the selected flame-retardants. Therefore fire safety should not be confused with non-flammability, and flame retardants only buy time in a real fire, as they inhibit ignition and propagation. As the fire progresses, the material (natural or synthetic polymer) may still burn.

There are many types of flame retardants, but they fall into a few main categories.

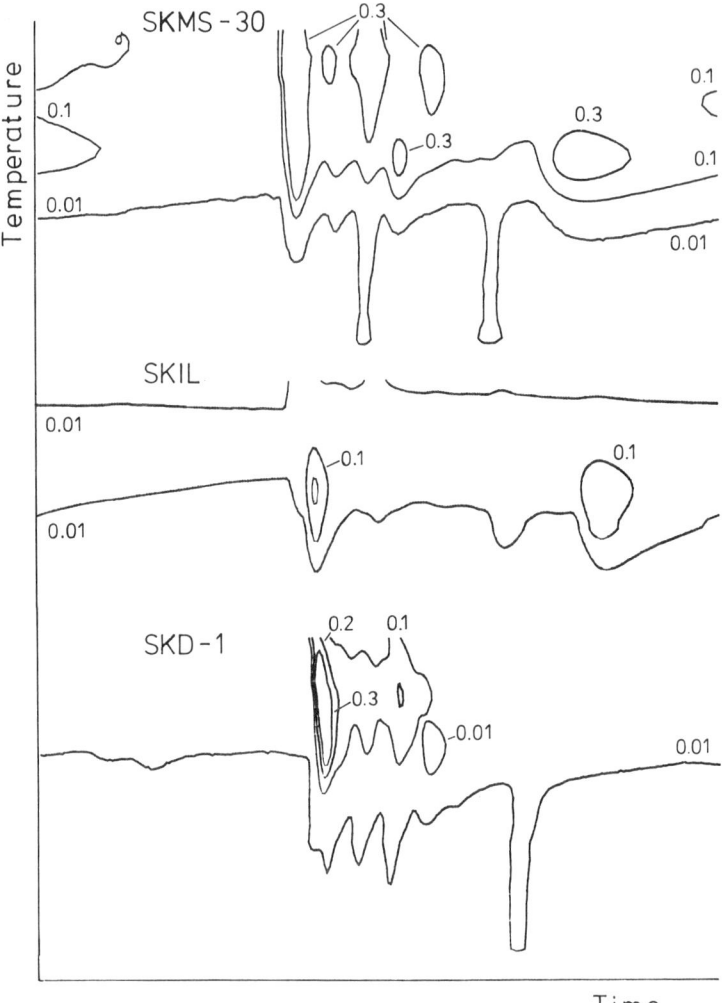

Fig. 5.25 — Contour plots of rubber thermochromatograms. Initial data are the same as for Figs. 5.23 and 5.24. The numbers beside the contours mark the cut height. The highest peak is taken as having a cut height equal to 1.0.

Extensive literature exists on the mechanism of thermal degradation of high-temperature resistant polymers and on the role of flame retardants [32].

Conventional pyrolysis gas chromatography, thermogravimetry and IR spectrometry are extensively used for studying the influence of flame retardants on the thermal stability of polymers and their resistance to fire. The main aim of these studies is determination of evolved gases, such as CO_2, CO, H_2O, H_2S and HCN. From this point of view, thermochromatography can give more precise information about the effect of flame retardants on the burning of polymers. These combustion products are detectable only by a thermal conductivity detector (HCN being an

Sec. 5.1]   **Characterization of gases evolved from polymers**   181

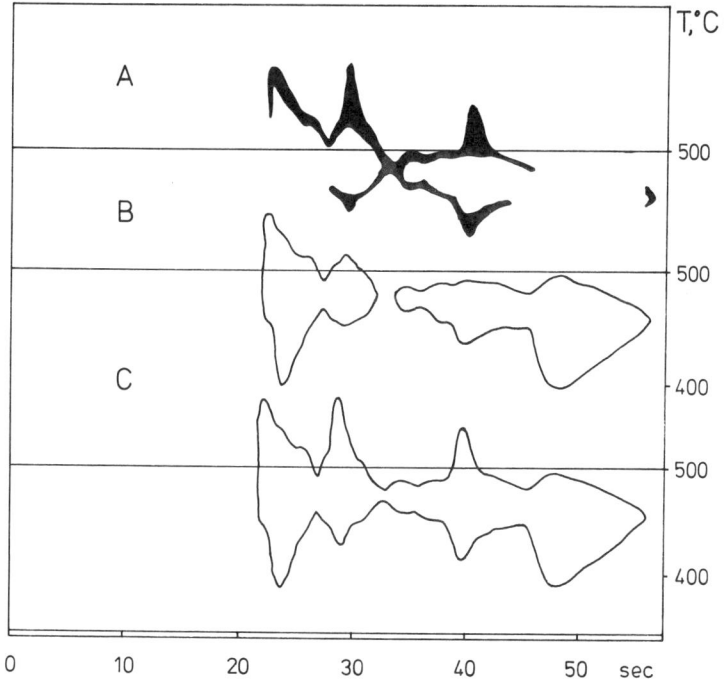

Fig. 5.26 — Contour plots of rubber thermochromatograms: C, 1,4-*cis*-polyisoprene with 5% 3,4-*cis*-polyisoprene; B, 1,4-*cis*-polyisoprene; A, contour pattern derived by subtraction of B from C.

exception), which is less sensitive than a flame ionization detector. With the correlation technique it is possible to observe small changes in the properties of polymers, that cannot be detected by other chromatographic techniques.

The advantages of thermochromatography in flame retardant studies are demonstrated in Fig. 5.29 where two sets of thermochromatogram slices show clearly the differences in the degradation mechanisms. Both cellulose fibres contain a phosphorus compound as a flame retardant. In A the cellulose fibres are crosslinked with phosphonitrilamide:

$R = -CH_2-CH(OH)-CH_2-$

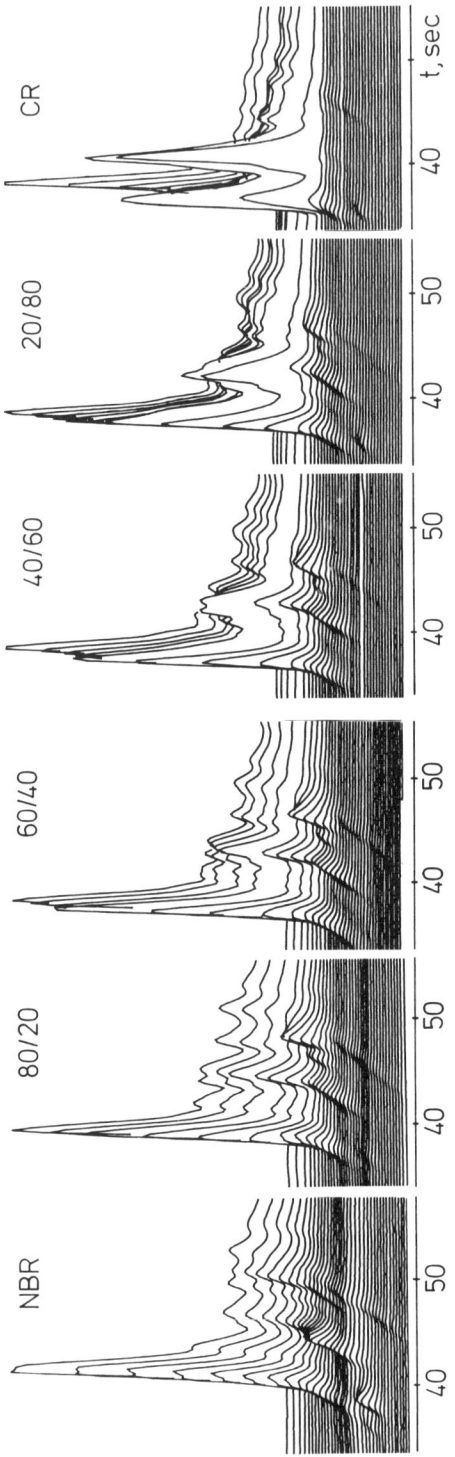

Fig. 5.27 — Stack plots of 100% butadiene–acrylonitrile copolymer (NBR) and 100% chloroprene (CR) and their mixtures. Temperature 250–450°C.

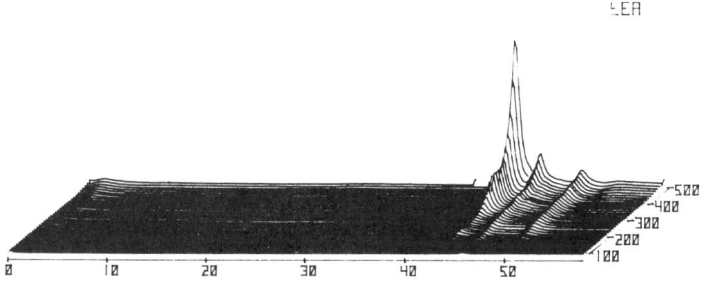
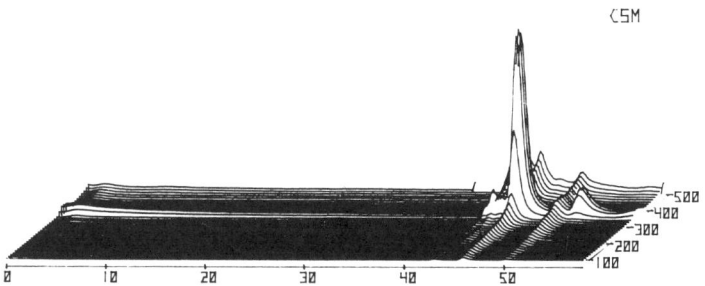
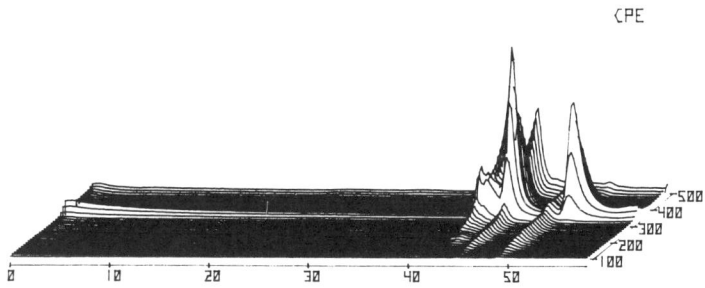

Fig. 5.28 — Thermal degradation of rubber-based insulating materials: EEA, 4 mg of ethene–ethyl acrylate copolymer; CSM, chlorosulphonated polyethylene; CPE, 4.4 mg of chlorinated polyethylene. Analysis conditions are the same as for Fig. 5.23.

and B is a copolymer of cellulose with poly(methylvinylpyridine metaphosphate).

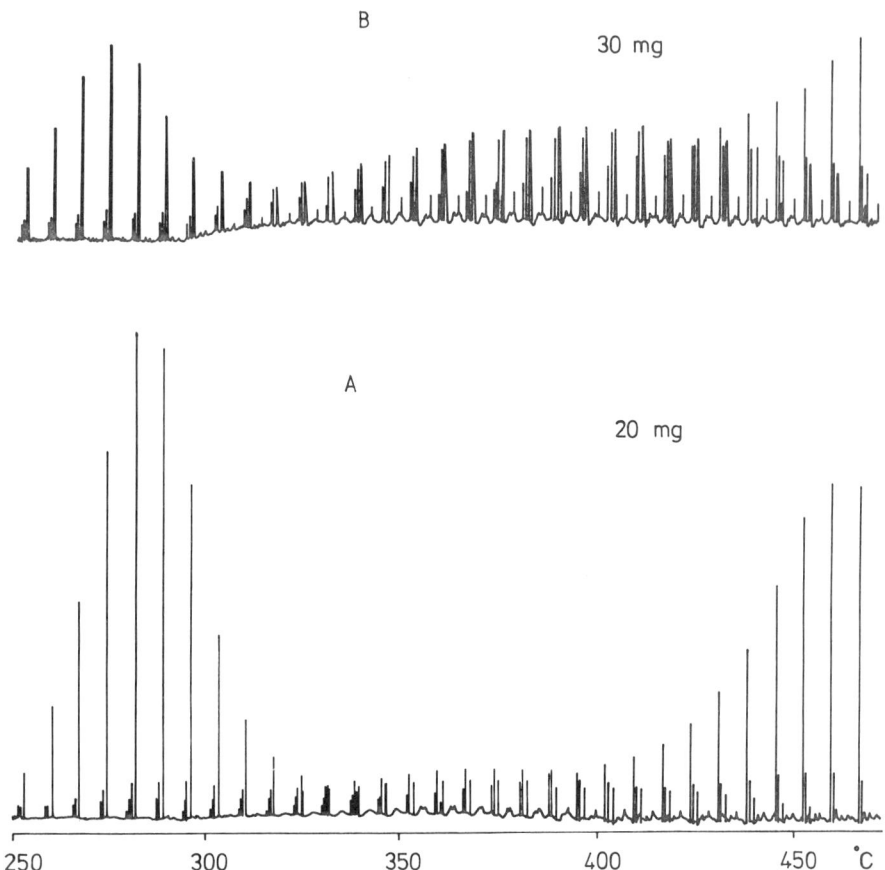

Fig. 5.29 — Thermal degradation of phosphorus-containing celluloses. A — contains phosphonitrilamide; B — contains poly(methlyvinylpyridine metaphosphate).

As can be seen, the degradation mechanisms are different. The methylvinylpyridine–cellulose copolymer is more thermally stable, yielding fewer evolved gases than does the phosphoronitrilamide. A systematic study of materials containing flame retardants by thermochromatography would produce valuable information for the building and furnishing industries.

5.1.3.6 Desorption of gases from solids
Thermochromatography, being a sensitive method, is very suitable for studying desorption processes.

The stack plots in Figs. 5.30 and 5.31 show the desorption of methanol and its degradation products from gallium and indium oxides after drying for 1 hr and 6 hr at 70°C. The baseline noise in the chromatograms is caused by the chromatograph being at maximum sensitivity because of the small amount of evolved gases. The

Sec. 5.1]   Characterization of gases evolved from polymers   185

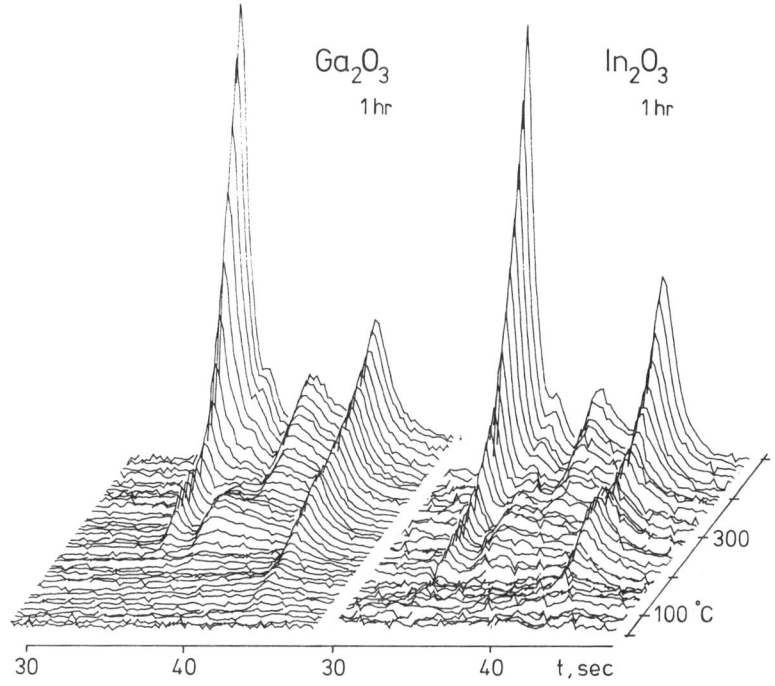

Fig. 5.30 — Desorption of methanol from gallium and indium oxides. The oxides were dried for 1 hr at 70°C. Sample weights 148.3 and 139.0 mg respectively.

sample size of the oxides is larger (110–150 mg) than that of rubber. For example, the sample size of the Ga_2O_3 was 148.3 mg. During the heating of the sample from 50 to 500°C in 35.7 min the weight loss was 1.38 mg, an average of 0.643 μg/sec. A flame ionization detector was used.

5.1.3.7 Study of the influence of catalysts on polymer thermal stability
Urea melamine formaldehyde resins are synthesized with use of different catalysts. The hazardous thermal degradation products and their evolution are of interest. On heating to 100–200°C, all three polymers evolve methanol and formaldehyde, but the amounts and ratio of these depend on the catalyst used. It can be clearly seen that $AlCl_3$, $FeCl_3$ and NH_4Cl influence the yields of the degradation products (Fig. 5.32).

5.1.3.8 Correlation gas chromatography in characterization of high-temperature resistant polymers
The advantage of correlation gas chromatography in studies of high-temperature resistant polymers has been demonstrated with a polyimide film [33]. Polyimide is a widely used thermally stable polymer. Although its critical thermal degradation begins at above 500°C, TG studies give evidence of some gas evolution at below this temperature [34]. The major degradation products are hydrogen, carbon monoxide,

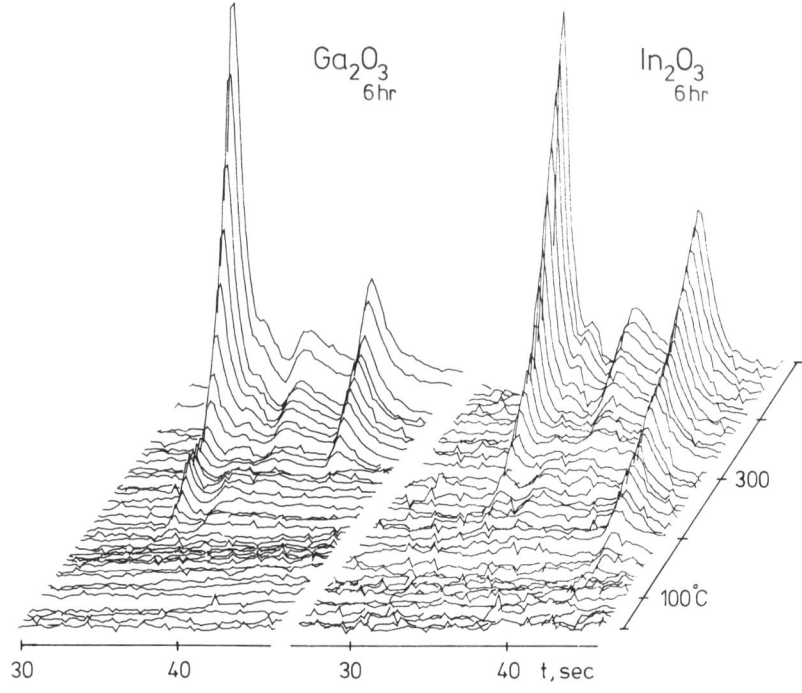

Fig. 5.31 — Desorption of methanol from gallium and indium oxides. The oxides were dried for 6 hr at 70°C. Sample weights 110.8 and 118.6 mg respectively.

carbon dioxide and water. To detect all these products a thermal conductivity detector is needed.

A typical chromatogram of the gases evolved from an industrial PM type polyimide film degraded at 300°C is given in Fig. 5.33. The experiment was continued for 3 hr and during that time 2.83 mg of degradation products passed through the detector, an average of 0.243 μg/sec. Because of the high thermal stability, the initial sample size was 75 mg, which is much higher than that used in conventional pyrolysis gas chromatography. The destruction of only a few mg of polyimide sample does not give a measurable signal in conventional GC. The evolution of the polyimide degradation products is presented in Fig. 5.34. As can be seen, there are two evolution maxima for water, at 60 and 220°C. The first maximum is due to moisture adsorbed on the polyimide surface and evaporating at 60°C. The second, at 220°C, is due to intermolecular water. Independent MS studies confirmed the gas evolution kinetics [35].

5.2 APPLICATIONS OF CORRELATION CHROMATOGRAPHY

5.2.1 Historical remarks

The idea of using a correlation technique in chromatography was proposed not by

Sec. 5.2] **Applications of correlation chromatography** 187

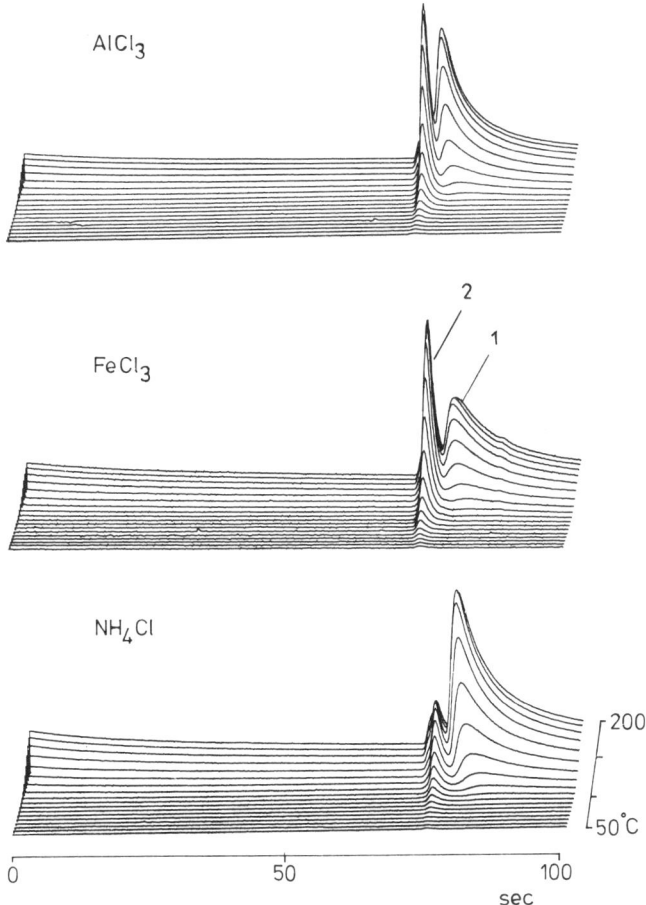

Fig. 5.32 — Degradation of urea melamine formaldehyde resins obtained by using different catalysts. Peaks: 1, methanol; 2, formaldehyde. Sample heating rate 5°C/min; GC column SE-30, oven temperature 130°C.

chromatographers but by specialists in the field of system identification (part of automatic control theory) [36–41]. These works are mainly theoretical studies describing the nature of the correlation technique. Ideas developed by the system identification theoreticians were realized experimentally by Smit [42], who first demonstrated the noise suppression properties of CC, and by Annino [43,44] who attempted to use CC for analyses of time-varying concentration flows. The first results were promising. The theory of CC was developed by Laurgeau [45,46], Annino [47], Phillips [48], Lub and Smit [49, 50] and by the authors of this book [51, 52] who proposed the FHT algorithm for CC [53].

A search of the literature has yielded several unusual applications of CC. Burke *et al.* used CC for adsorption studies [54] and also frequency modulated CC (Section

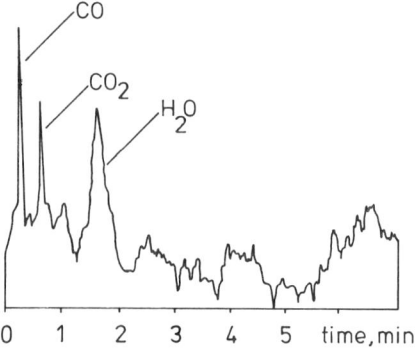

Fig. 5.33 — Correlogram of degradation products of polyimide evolved at a temperature of 300°C. (Reprinted with permission, from M. Kaljurand and E. Küllik, *Trends Anal. Chem.*, 1985, **4**, 200. Copyright 1985, Elsevier Science Publishers).

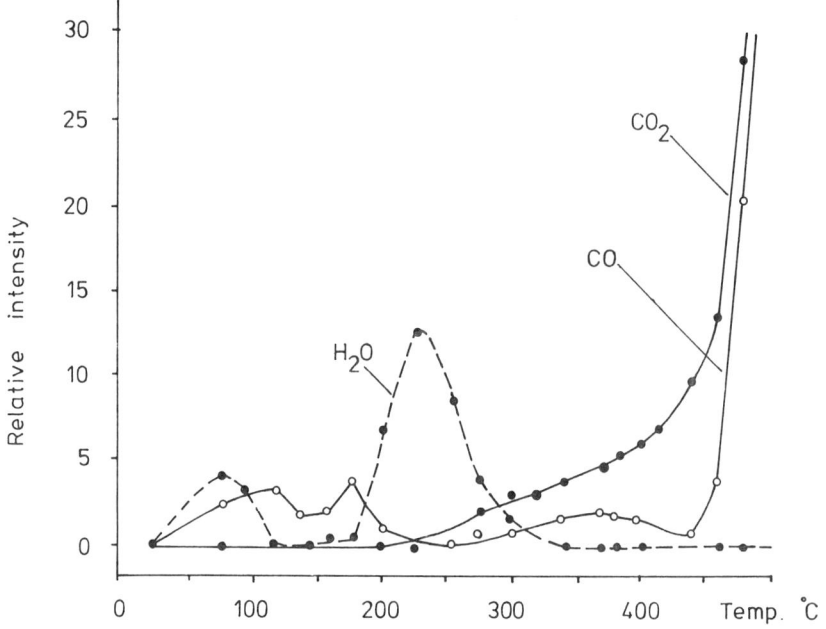

Fig. 5.34 — Evolution rate of polyimide degradation products. (Reprinted with permission, from M. Kaljurand and E. Küllik, *Trends Anal. Chem.*, 1985, **4**, 200. Copyright 1985, Elsevier Science Publishers).

1.2.2) [55]. Laurgeau used CC for simultaneous measurement of two independent flows [56]. The advantages and potentials of the technique were discussed in [57] and simultaneously CC was applied experimentally in a calibration procedure in HPLC. Kaljurand et al. started the study of the composition of the degradation products of high molecular-weight compounds [58, 59]. Frazer studied the possibilities of CC in time-varying process analysis [60]. However ingenious these methods seem at first sight, the success of these early applications is doubtful. The main reason for saying this is that the influence of correlation noise was underestimated. Theoretical studies and simulations on computers showed the great importance of the correlation noise in CC measurements and thus the importance of improving the performance of the correlation chromatographic equipment. This was demonstrated in Section 2.2. The efforts of investigators are now being directed to that end. The main target for improvement are the input systems, which are the weakest part of CC. The other parts, the chromatography and the data-processing and control devices are sufficiently developed today, but the input systems are still far from ideal. The necessity of performing thousands of reproducible injections during one correlation experiment, with a frequency of 0.5–1 injections per second, presents a completely new set of requirements for the sampling system. Mechanical valves are certainly not well suited to the task and other possibilities for sample introduction have been extensively investigated. Pneumatic flow switches (Section 4.3.2) are attractive because of the minimum mechanical movement involved, and flow modulators (Section 4.3.1), which do not contain any mechanical parts at all, are especially useful. Modulators for correlation chromatography were developed by Phillips and Valentin [61–68].

The recent applications of CC are mainly concerned with environmental problems. As we saw in Section 2.3, CC has several advantages over the the standard procedures of environmental chemistry for the estimation of compounds at low concentrations in large samples. Below, we consider some applications of CC in determining impurities in air and water.

5.2.2 Application of CC in gas analysis

The first applications of correlation gas chromatography in determination of impurities in light gas mixtures were mainly demonstrations of the multiplex advantages (i.e., measurement at below noise level) rather than serious applications of the technique. Moss and Godfrey determined methane in air [69, 70], Annino determined ethene and propene in nitrogen [71] and hydrogen and oxygen in helium [72]. Koel et al. determined methane in helium [73]. In that work the system was calibrated at below the noise level by using exponential dilution of the sample for low-concentration flow generation. The amount detected in these experiments was about 10 pg when a flame ionization detector was used. This quantity (depending on the column operating parameters) gives a chromatographic peak that is at or just below the detector noise level when introduced in a single injection. However, 10 pg is not the lower limit of determination with an FID. This limit led to criticism of correlation chromatography and to disappointment in it for some of the investigators [72,74]. However, it seems to have been overlooked by some critics that CC deals with the untreated sample whilst most methods of low-concentration measurement require some sample pretreatment. This gives an advantage to CC in field measurements (e.g. on board a research plane or ship). In the measurement of small trace amounts in a large amount of matrix it is advantageous to use the pure matrix as the carrier gas, otherwise any fluctuations in the matrix peaks generate correlation noise.

Using CC with differential chromatography (Section 1.2.1) reduces the possible high background signal that will be present if only one column and detector are used with the matrix compound as a carrier. The idea of differential CC was proposed in [75]. This technique may be refined even more, by using specially developed filters to block out several target compounds in the reference path of the differential chromatograph and thus to form a 'pure' reference carrier. In the differential two-detector system the components having the same concentration in the analytical and reference flows will not give a signal. The components that are blocked in the reference flow by a specific filter will appear in the analytical flow and give a signal in the detector. It is an extremely sensitive method for specific determinations. An extensive list of filter materials is given in [76].

Methane is of interest in the study of the earth's atmosphere because of the implications of the 'greenhouse effect'. This effect is usually attributed to an increase in carbon dioxide but it is expected that methane will also contribute to it. The technique of removing all unwanted compounds by special filters was used for methane determination in ambient air [77]. The authors, however, did not use a differential scheme but only a simple one-column and detector chromatograph. The chromatographic system developed for this work consisted of a compressor that supplied the ambient air for the measurement and a drier that removed all trace compounds from the air flow except the noble gases, nitrogen, oxygen, hydrogen, carbon monoxide, carbon dioxide and methane. The air flow containing these compounds was fed through a modulator to a column packed with Alcoa Type F-1 alumina (80–100 mesh). The modulator was a 10-cm long stainless-steel tube filled with silver(I) oxide powder heated to 230°C. The hot silver oxide reacts catalytically with methane and hence removes it from the gas flow. The other compounds that pass through the drier were found to be unaffected by the modulator. Modulation was accomplished by cooling the bed with pulses of an external stream of compressed air, thus momentarily allowing methane in the air flow to reach the column. The efficiency of this modulation was low (estimated to be 1%) but a sufficiently large signal was produced to allow determination of ambient methane concentrations. The methane concentration impulse passed through the column and was detected by a tin–metal oxide semiconductor detector that operates by catalytically oxidizing hydrocarbons in the presence of air and measuring the heat released.

In multiplex experiments this detector had a detection limit of 1 ppm v/v for methane [77]. This detection limit is comparable to that of conventional chromatography, showing there is no multiplex advantage over the common method. However, because of the low efficiency of the modulator the amount of methane injected by one pulse in the multiplex experiments (strictly speaking the concentration perturbation relative to the mean methane flow in air) is far less than the quantity of methane used for determination of the detection limits in an ordinary GC with a 1-ml injection and a tin–metal oxide semiconductor detector. Thus increasing the modulator efficiency will decrease the detection limit in these experiments. On the other hand, the main feature of chromatography, the separation of compounds, was not used in this work at all, because only one compound was present. However, for future work it will be important to determine several inorganic trace compounds

Sec. 5.2]   Applications of correlation chromatography   191

in the air simultaneously and the separation efficiency will then become important. Thus, although the multiplex advantage was not fully exploited in [77] the potential for its use was inherent in the method. The authors determined the ambient methane concentration at hourly intervals over an 8-day period (Fig. 5.35), near San Francisco

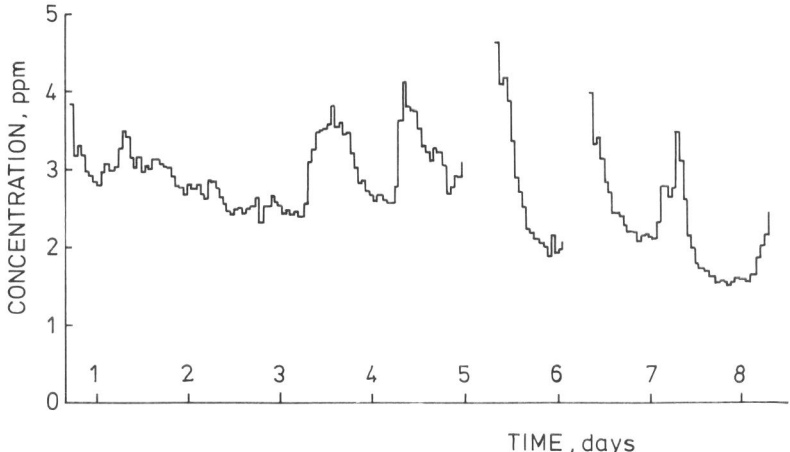

Fig. 5.35 — Profile of the concentration of methane in ambient air, showing 1-hr averages for an 8-day period (San Francisco Bay area, May 1984). Some data points are missing because of computer malfunctions. (Reprinted with permission from J. R. Valentin, G. C. Carle and J. R. Phillips, *Anal. Chem.*, 1985, **57**, 1035. Copyright 1985, American Chemical Society).

Bay. The lowest concentration of methane found was 1.5 ppm v/v, which is comparable to the reported values, indicating that the method gave reasonable results. The data showed evidence of some regular variations, but it was not determined whether the sources of these perturbations were man-made, natural, or both, as there are many potential methane sources in the vicinity of a highly urban area.

Phillips *et al*. [78] obtained multiplex chromatograms by using hydrogen to carry p-di-isopropylbenzene at 7.6×10^{-16} g/sec through a 20-cm thermal desorption modulator followed by a 9-m SE-52 column. The modulator was made from a silica capillary tube coated with electrically conductive paint. The total modulator signal duration was 3840 sec, made up of a sequence of random (not PRBS) pulses with an average time of 8 sec between pulses. Thus, about 480 impulses were used. The correlogram signal was slightly above the detection limit. The result was about two orders of magnitude below the concentration detection limit for conventional single-injection determination with an FID. This value, the lowest on record, can be only partly attributed to the multiplex advantage, because (Section 2.2.3) the signal-to-noise ratio should be improved by a factor of $\sqrt{480}/2 = 15.5$. On the other hand the definition of the detection limit and the signal-to-noise ratio is quite fuzzy in chromatography, and different authors use different definitions for these quantities.

The reason for this good result is partly the performance of the short capillary column, where the compound bands are very narrow and hence concentrated, and give a large response in the detector, and partly the performance of the modulator itself, which enables the introduction of larger amounts of sample than would be possible with a single injection.

Measurements at lower concentration levels are complicated by memory effects from previous samples and from contaminants in the carrier gas. The chromatograph must be 'cleaned' of the unwanted sources of the sample, including bleeding from the septum and/or from the walls of the sampling system. Phillips reported the appearance of 'ghost' peaks with an amplitude much larger than that for the p-di-isopropylbenzene itself.

Correlation chromatography has been applied to analysis of the gaseous products from a polyimide film [33,79]. The determination of CO, CO_2, and H_2O has already been described in Section 5.1.3.8.

An interesting potential application of multiplex gas chromatography as an analytical technique for planetary atmosphere studies has been reported [80,81]. These studies, made on board a spacecraft, would be a challenge to instrument designers because of the need for advanced instrumentation which would be highly sensitive, efficient, physically small and capable of analysing complex mixtures [82]. Multiple injection together with chemical modulators offers several advantages over conventional chromatography in this kind of study, offering improvement in detection limits, the analysis of complex mixtures by selective modulation of some components of the sample, and reduction of the amount of expendables needed to run an analysis.

5.2.3 Application of CC in water analysis

In water analysis by GC it is convenient to use water both as a carrier gas and as a liquid phase. This is steam–solid chromatography. Water does not give a response in a flame ionization detector, but increases the detector noise and decreases the sensitivity, making it difficult to detect impurities in water by direct injection of aqueous samples and analysis with an FID. Koel [83] used steam–solid chromatography and CC for the determination of alcohols and phenol in water samples (Fig. 5.36). Samples were transferred into the steam flow and a pneumatic Deans-type sampling valve was switched between the pure steam and the sample steam flows. The detection limits obtained were 1 part in 10^7 v/v (10^{-4} g/l.) for phenol and 1 part in 10^9 v/v (10^{-6} g/l.) for alcohols, values lower than it is generally possible to obtain in steam–solid chromatography with a single injection. Koel used glass columns filled with inert materials (Porasil and Spherosil). Filling was necessary to ensure a large surface area for condensing steam to form a liquid phase for partitioning of the components. Thus a water sample passes through the system without interference or reaction. This is an important feature of this kind of experiment if a reacting species in water is of interest. Also the water in any active adsorption sites should block steam–solid chromatography. This system should be a good prototype for a field instrument on a research ship if the flame ionization detector could be replaced by some simpler device (e.g. a thermal conductivity detector); the instrument would then need no consumables except electric power. The possibly higher detection limit (compared to that of the FID-based instrument) would be compensated for by the multiplex advantage.

Sec. 5.2] **Applications of correlation chromatography** 193

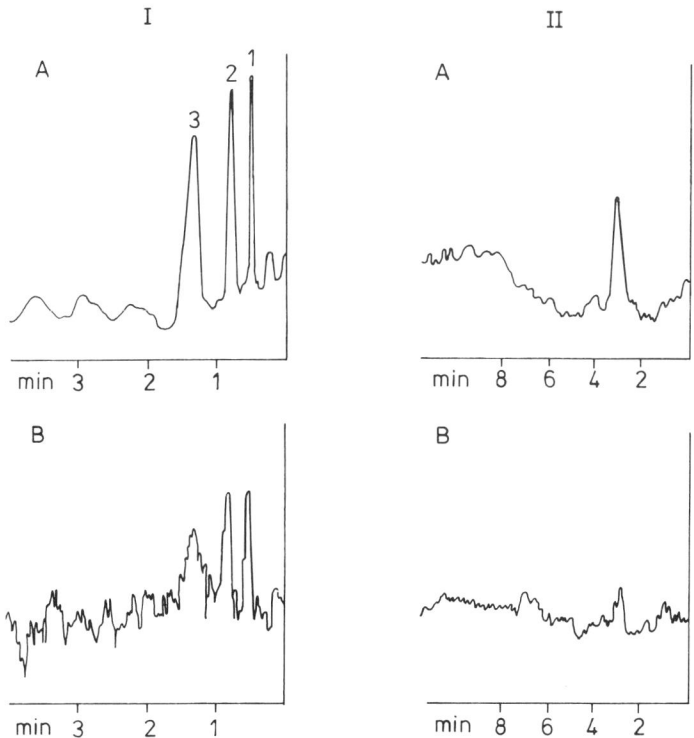

Fig. 5.36 — Correlograms (A) and single-injection chromatograms (B); I — alcohols (5 μg/l.): 1, ethanol; 2, n-propanol; 3, n-butanol; II — phenol (0.7 mg/l.); steam–solid chromatography with flame ionization detection. (Adapted from [83]).

In a series of publications Smit *et al.* described the apparatus and methodology for using the correlation technique in high-performance liquid chromatography [84–91]. In HPLC the detection is generally more of a problem than the injection. The main topic of CC–HPLC in Smit's work was analysis for phenol derivatives. The detection limit was lowered by two orders of magnitude. To achieve this improvement the total time of the CC experiment was 2 hr. The detection limit for the phenols and chlorinated phenols with a UV detector was 6 μg/l. at 254 nm.

The analytical performance of CC was demonstrated by extending the calibration graph for phenol with data found by CC. The total range of phenol concentration measured was four orders of magnitude from 0.01 to 100 μg/l. The detection limit achieved with the fluorimetric detector was 3 ng/l.

The application of CC in HPLC demanded the development of a new injection device instead of the commonly used mechanical valves, which were found to be subject to adsorption effects and to corrosion by the eluent. The authors developed a new sample reservoir/injection device that consisted of an 8.5 cm × 40 mm i.d. stainless-steel tube. One end of the reservoir was the water inlet. Inside the reservoir was a precision glass tube (65 mm long × 30 mm i.d.) that was used as a sample reservoir, one end of the reservoir opening into the HPLC column and the other sealed with a polytetrafluoroethylene plunger. A pump forced water into the system of the two reservoirs, one containing pure eluent and the other the sample. Pressure was applied to the sample plunger or the eluent plunger through electromagnetic valves. The inlet system is presented in Fig. 5.37.

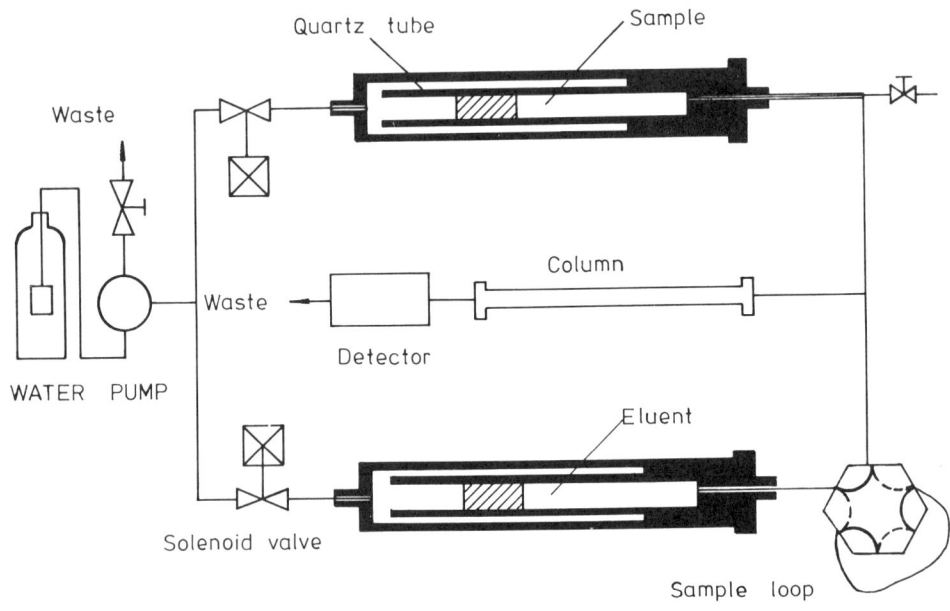

Fig. 5.37 — Sample holders and injection system for correlation HPLC (simplified). (Reprinted with permission from H. C. Smit, T. T. Lub and W. J. Vloon, *Anal. Chim. Acta*, 1980, **122**, 267. Copyright 1980, Elsevier Science Publishers).

A typical chromatogram of chlorinated phenols thus obtained is shown in Fig. 5.38. This figure demonstrates that the complexity of the sample is not a problem in CC. The first peak is probably due to the eluent, tetrahydrofuran, and its oxidation products. Two more peaks appear in the correlogram than in the chromatogram. The small peak in the middle is probably for an isomer. The second large peak may be a late eluting peak from a previous injection, or a peak that is 'folded back' because the duration of the PRBS is shorter than the chromatogram, a typical CC phenomenon. However, in the latter case the origin of the peak is not clear. A probable explanation is that some pollution originating from the eluent has given rise to a peak somewhere after the last peak for the sample [90].

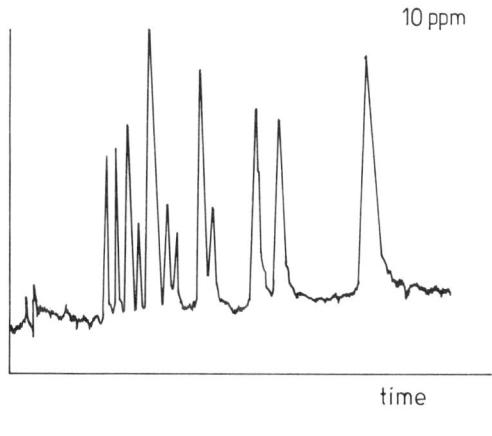

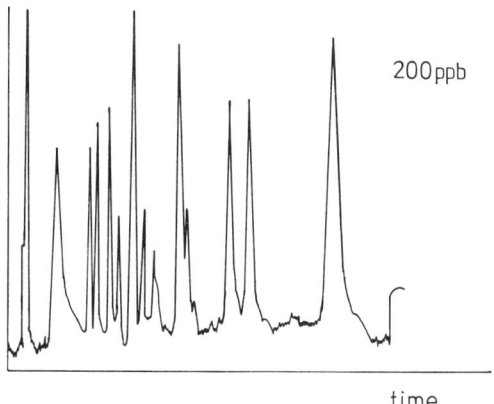

Fig. 5.38 — Separation of different chlorinated phenols by correlation chromatography (200 ppb) and conventional HPLC (10 ppm). For identification and chromatographic details see [102]. (Reprinted with permission from H. C. Smit, T. T. Lub and W. J. Vloon, *Anal. Chim. Acta*, 1980, **122**, 267. Copyright 1980, Elsevier Science Publishers).

REFERENCES

[1] G. H. Langer, *Thermochim. Acta*, 1986, **100**, 187.
[2] H. L. C. Meuzelaar, J. Haverkamp and F. D. Hileman, *Techniques and Instrumentation in Analytical Chemistry, Vol. 3*, Elsevier, Amsterdam, 1982.
[3] T. H. Risby and J. A. Yergey, *Anal. Chem.*, 1982, **54**, 2228.
[4] J. F. Holland, C. G. Enke, J. Allison, J. T. Stults, J. D. Pinkston, B. Newcome and J. F. Watson, *Anal. Chem.*, 1983, **55**, 997A.
[5] W. J. Irwin, *Analytical Pyrolysis*, pp. 121–235. Decker, New York, (1982).
[6] P. A. Hmelnitsky, I. M. Lukasenko and E. S. Brodsky, *Pyrolysis Mass Spectrometry for High Molecular Compounds*, Khimiya, Moscow, 1980, in Russian.
[7] G., R. Davidson and T. G. Mathys, *Anal. Chem.*, 1986, **58**, 837.
[8] M. C. Earnest, *Anal. Chem.*, 1984, **56**, 1471A.
[9] G. Blandenet, *Chromatographia*, 1969, **2**, 184.
[10] F. Zitomer, *Anal. Chem.*, 1968, **40**, 1091.
[11] F. O. Rice and K. K. Rice, *The Aliphatic Free Radicals*, Johns Hopkins Press, Baltimore, 1968.
[12] S. L. Madorsky and S. Straus, *J. Res. Natl. Bur. Stds.*, 1948, **41**, 315.
[13] L. A. Wall, *J. Res. Natl. Bur. Stds.*, 1948, **41**, 315.

[14] R. B. Barnes, R. C. Gore, R. W. Stafford and V. Z. Williams, *Anal. Chem.*, 1948, **20**, 402.
[15] A. T. James and A. J. P. Martin, *Analyst*, 1952, **77**, 915.
[16] W. H. T. Davidson, S. Slaney and A. L. Wragg, *Chem. Ind.*, 1954, 1356.
[17] E. A. Radell and H. C. Strutz, *Anal. Chem.*, 1959, **31**, 1890.
[18] R. L. Lehre and J. C. Robb, *Nature*, 1959, **183**, 1671.
[19] O. D. Hummel and F. Scholl, *Atlas of Polymer and Plastic Analysis*, Vol. 1, Hanser Verlag, Munich, 1987.
[20] R. S. Lehre and J. C. Robb, *J. Gas Chromatog.*, 1967, 89.
[21] Y. Sugimura, S. Tsuge and T. Takeuchi, *Anal. Chem.*, 1978, **50**, 1172.
[22] I. Ericsson, *J. Chromatog. Sci.*, 1978, **16**, 340.
[23] J. C.-A. Hu, *Anal. Chem.*, 1981, **53**, 311A; 794A.
[24] P. R. Lattimer, *Anal. Chem.*, 1981, **53**, 794A.
[25] S. A. Liebman, D. H. Ahlstrom and C. R. Foltz, *J. Polym. Sci., Polymer Chem.*, 1978, **16**, 3139.
[26] K. Yamada, S. Oura and T. Hasuki, *Proc. IV Int. Conf. Thermal Analysis*, Budapest, 1974, **3**, 1029.
[27] C. C. Williams, *J. Chem. Soc.*, 1862, **15**, 110.
[28] M. G. Bouchardat, *Bull. Soc. Chim. France*, 1875, **24**, 108.
[29] T. Midgley Jr. and A. L. Henne, *J. Am. Chem. Soc.*, 1929, **51**, 125.
[30] A. Krishen, *Anal. Chem.*, 1972, **44**, 494.
[31] C. H. Bamford and C. F. H. Tipper, *Comprehensive Chemical Kinetics*, Vol. 14, p. 44. Elsevier, Amsterdam, 1975..
[32] A. H. Frazer, *High Temperature Resistant Polymers*, Wiley–Interscience, New York.
[33] M. Kaljurand and E. Küllik, *Trends Anal. Chem.*, 1985, **4**, 200.
[34] J. A. Hider, *Lab. Prac.*, 1986, **35**, No. 2, 19.
[35] A. S. Teleshova, E. N. Teleshov and A. N. Pravdennikov, *Visokomol. Soeding., Ser. A.*, 1975, **17**, 134.
[36] K. Izawa, K. Furuta, T. Fujiwara and N. Suyawa, *Ind. Chim., Belge*, 1967, **32**, 223.
[37] W. D. T. Davies, *Instrum. Pract.*, 1968, **22**, 213.
[38] P. Eykoff, *System Identification–Parameter and State Estimation*, Wiley, New York, 1974.
[39] K. R. Godfrey and Devenish, *Meas. Contr.*, 1969, **2**, 228.
[40] R. Hoffmann, M. M. Gupta and P. N. Nikiforuk, *Proc. IEE*, 1972, **119**, 237.
[41] D. R. Owens, *Chem. Brit.*, 1972, **8**, 469.
[42] H. C. Smit, *Chromatographia*, 1970, **3**, 515.
[43] R. Annino and L. Bullock, *Gas Chromatography 1972*, S. G. Perry and E. R. Adlard (eds.), p.171. Appl. Sci. Publ., London, 1972.
[44] R. Annino and L. Bullock, *Anal. Chem.*, 1973, **45**, 1221.
[45] C. Laurgeau and B. Espiau, *J. Chim. Phys.*, 1974, **71**, 1143.
[46] C. Laurgeau and F. Barras, *Chromatographia*, 1975, **8**, 373.
[47] R. Annino and E. Grushka, *J. Chromatog. Sci.*, 1976, **14**, 265.
[48] J. B. Phillips, *Anal. Chem.*, 1980, **52**, 468A.
[49] T. T. Lub and H. C. Smit, *Anal. Chim. Acta*, 1979, **112**, 341.
[50] T. T. Lub, *Correlation Chromatography*, Dissertation, p. 85. Univ. Amsterdam, 1980.
[51] M. Kaljurand and E. Küllik, *J. Chromatog.*, 1979, **186**, 145.
[52] E. Küllik and M. Kaljurand, *Multiplex Measurement Method in Chromatography*, in *Summaries in Science: Chromatography*, Vol. 4, p. 3, VINITI, Moscow, 1983.
[53] M. Kaljurand and E. Küllik, *Chromatographia*, 1970, **11**, 328.
[54] J. B. Phillips and M. F. Burke, *J. Chromatog. Sci.*, 1979, **14**, 495.
[55] D. Villalanti, M. F. Burke and J. B. Phillips, *Anal. Chem.*, 1979, **51**, 2222.
[56] C. Laurgeau and F. Barras, *Chromatographia*, 1979, **12**, 160.
[57] H. C. Smit, C. Mars and J. C. Kraak, *Anal. Chim. Acta*, 1986, **181**, 37.
[58] M. Kaljurand and E. Küllik, *J. Chromatog.*, 1979, **171**, 243.
[59] E. Urbas, M. Kaljurand and E. Küllik, *J. Anal. Appl. Pyrolysis*, 1980, **1**, 231.
[60] S. R. Frazer, *Information Extraction in Chromatography Using Correlation Techniques*, Dissertation, Univ. Arizona, 1985.
[61] D. Carney and J. B. Phillips, *HRC&CC*, 1981, **4**, 413.
[62] J. R. Valentin, G. C. Carle and J. B. Phillips, *HRC&CC*, 1982, **5**, 269.
[63] J. R. Valentin, G. C. Carle and J. B. Phillips, *HRC&CC*, 1983, **6**, 621.
[64] J. B. Phillips, J. R. Valentin and G. C. Carle, *Toxic Materials in the Atmosphere*, p. 135. ASTM STP 786, 1982.
[65] J. B. Phillips, *Trends Anal. Chem.*, 1982, **1**, 163.
[66] K. Jinno, D. P. Carney and J. B. Phillips, *Anal. Chem.*, 1986, **58**, 1248.
[67] D. P. Carney and J. B. Phillips, *Anal. Chem.*, 1986, **58**, 1251.

[68] J. B. Phillips, D. Luu and R. Lee, *J. Chromatog. Sci.*, 1986, **24**, 396.
[69] G. C. Moss and K. R. Godfrey, *Instrum. Technol.*, 1973, **20**, 33.
[70] G. C. Moss, P. J. Kipping and K. R. Godfrey, *Int. Gas Chromatography 1972*, S. G. Perry and E. R. Adlard (eds.), p. 187. Appl. Sci. Publ., London, 1973.
[71] R. Annino, M. F. Gonnord and G. Guiochon, *Anal. Chem.*, 1979, **51**, 379.
[72] R. Annino and J. Leone, *J. Chromatog. Sci.*, 1982, **20**, 19.
[73] M. Koel, M. Kaljurand and E. Küllik, *Proc. Estonian SSR Acad. Sci., Chem.*, 1983, **32**, 125.
[74] R. Annino, in *Trace Residue Analysis*, D. A. Kurtz (ed.), p. 83. ACS, Washington DC, 1985.
[75] H. C. Smit, T. T. Lub and R. P. Duursma, *27th IUPAC Congress*, J. Larinkari and J. Oksanen (eds.), Abstract No. 614.
[76] J. Drugov and V. Berezkin, *Gas Chromatographic Analysis of Polluted Air*, Khimiya, Moscow, 1981 (in Russian).
[77] J. R. Valentin, G. C. Carle and J. B. Phillips, *Anal. Chem.*, 1985, **57**, 1035.
[78] J. B. Phillips, D. Luu, J. B. Pawliszyn and G. C. Carle, *Anal. Chem.*, 1985, **57**, 2779.
[79] E. Küllik and M. Kaljurand, *Anal. Chim. Acta*, 1986, **181**, 51.
[80] J. R. Valentin, G. Carle and J. B. Phillips, in *2nd Symposium on Chemical Evolution and Origin of Life*, Abstracts, p. 33. NASA Ames Res. Center, Moffet Field, CA, USA 1985.
[81] J. R. Valentin, *Multiplex Gas Chromatography for Use in Space Craft*, NASA Techn. Memor. 86668, NASA Ames Res. Center, Moffet Field, CA, USA, 1985.
[82] R. J. Simpson, *Trends Anal. Chem.*, 1985, **4**, 138.
[83] M. Koel, *Proc. Estonian SSR Acad. Sci., Chem.*, 1986, **35**, 154.
[84] T. T. Lub, H. C. Smit and H. Poppe, *J. Chromatog.*, 1978, **149**, 771.
[85] H. Smit, *Anal. Chim. Acta*, 1980, **122**, 201.
[86] H. Smit, T. T. Lub and W. J. Vloon, *Anal. Chim. Acta*, 1980, **122**, 267.
[87] H. C. Smit, R. P. J. Duursma and H. Steizstra, *Anal. Chim. Acta*, 1981, **133**, 283.
[88] J. M. Laeven, H. C. Smit and J. C. Kraak, *Anal. Chim. Acta*, 1983, **150**, 253.
[89] H. C. Smit, *Trends Anal. Chem.*, 1983, **2**, 1.
[90] H. C. Smit, in *Trace Residue Analysis*, D. A. Kurtz (ed.), p. 101. ACS, Washington DC, 1985.
[91] H. C. Smit, *Data Analysis Chromatography*, in *Chemometrics, Mathematics and Statistics in Chemistry*, B. R. Kowalski (ed.), p.225. Reidel, Dordrecht, Holland, 1984.

Appendix 1
Some matrix algebra

Here we give some basic definitions of the terms and the operations in matrix algebra used in this book. For greater details see [1,2].

A matrix **A** is a table of numbers having n rows and m columns. If $m = n$ we talk about a square matrix. If $m = 1$ we are talking about a vector. The numbers (A_{ij}) that compose a matrix are called matrix elements. The index i is the row index ($i = 1, 2, \ldots, n$) and the index j is the column index ($j = 1, 2, \ldots, m$).

Examples: Let $n = 2$ and $m = 3$. A 2×3 matrix with elements A_{ij} can be written as follows:

$$\mathbf{A} = \begin{bmatrix} A_{11} & A_{12} & A_{13} \\ A_{21} & A_{22} & A_{23} \end{bmatrix}$$

This matrix can be considered to be composed of three column vectors.

$$C_1 = \begin{bmatrix} A_{11} \\ A_{21} \end{bmatrix}; \quad C_2 = \begin{bmatrix} A_{12} \\ A_{22} \end{bmatrix}; \quad C_3 = \begin{bmatrix} A_{13} \\ A_{23} \end{bmatrix}$$

A transpose of a matrix, $\mathbf{A}^T$ is obtained from **A** by transposing the rows and columns, i.e.

$$\mathbf{A}^T = \begin{bmatrix} A_{11} & A_{21} \\ A_{12} & A_{22} \\ A_{13} & A_{23} \end{bmatrix}$$

Thus if a vector C is a one-column matrix, the vector C^T is a one-row matrix.

Addition and substraction of two matrices is defined as addition and substraction of the corresponding elements, i.e. if

$$\mathbf{A} = \begin{bmatrix} A_{11} & A_{12} \\ A_{21} & A_{22} \end{bmatrix}; \quad \mathbf{B} = \begin{bmatrix} B_{11} & B_{12} \\ B_{21} & B_{22} \end{bmatrix}$$

Then

$$\mathbf{A} \pm \mathbf{B} = \begin{bmatrix} A_{11} \pm B_{11} & A_{12} \pm B_{12} \\ A_{21} \pm B_{21} & A_{22} \pm B_{22} \end{bmatrix}$$

For addition or substraction, the number of rows and columns of matrix **A** must be equal to the number of rows and columns of matrix **B**.

Multiplication of a matrix **A** by a number s (a scalar) is defined as the multiplication of each matrix element by s:

$$s\mathbf{A} = \begin{bmatrix} sA_{11} & sA_{12} \\ sA_{21} & sA_{22} \end{bmatrix}$$

Multiplication of two matrices **A** and **B** is defined as follows: Let $\mathbf{C} = \mathbf{AB}$. An element of **C**, $C_{ij} = \sum_{k=1}^{n} A_{ik} B_{kj}$ where n is the number of columns in **A** and the number of rows in **B**, is produced by multiplying the elements of row **A** by the elements of a column of **B** and summing. It follows from this definition that if **A** is an $m_1 \times n_1$ matrix and **B** is an $m_2 \times n_2$ matrix, then it is possible to multiply **A** by **B** only if $m_2 = n_1$.

Example: If

$$\mathbf{A} = \begin{bmatrix} A_{11} & A_{12} & A_{13} \\ A_{21} & A_{22} & A_{23} \end{bmatrix}; \quad \mathbf{B} = \begin{bmatrix} B_{11} & B_{12} \\ B_{21} & B_{22} \\ B_{31} & B_{32} \end{bmatrix}$$

then

$$\mathbf{C} = \mathbf{AB} = \begin{bmatrix} A_{11}B_{11} + A_{12}B_{21} + A_{13}B_{31} & A_{11}B_{12} + A_{12}B_{22} + A_{13}B_{32} \\ A_{21}B_{11} + A_{22}B_{21} + A_{23}B_{31} & A_{21}B_{12} + A_{22}B_{22} + A_{23}B_{32} \end{bmatrix}$$

The matrix multiplication definition arises from the requirement to find a compact writing form for the operations of linear algebra. For example, a set of linear equations can be written easily as multiplication of a matrix **A** by a vector of unknowns *X* as $\mathbf{A}X = B$, where *B* is a column vector. Matrix multiplication has all the properties of multiplication of numbers (scalars) except the commutative property (i.e. in general, $\mathbf{AB} \neq \mathbf{BA}$ except for certain types of matrices). Also $(\mathbf{AB})^T = \mathbf{B}^T \mathbf{A}^T$ and $\mathbf{AA} = \mathbf{A}^2$.

Special matrices
1. A unity (or identity) matrix (**I**) is a matrix that has ones in the main diagonal and zeros elsewhere:

$$\mathbf{I} = \begin{bmatrix} 1 & 0 & \cdots & 0 \\ 0 & 1 & & 0 \\ \vdots & & \ddots & \vdots \\ 0 & 0 & \cdots & 1 \end{bmatrix}$$

Multiplying a matrix by the unity matrix results in the same matrix: $\mathbf{IA} = \mathbf{AI} = \mathbf{A}$.

2. An inverse matrix $(\mathbf{A}^{-1})$ is defined through the relation $\mathbf{A}^{-1}\mathbf{A} = \mathbf{I}$. Not every matrix has an inverse and it is usually not easy to find an inverse matrix.

3. A permutation matrix $(\mathbf{P})$ is a matrix that changes the order of the rows or the columns of a matrix. If $\mathbf{A}$ is multiplied from the left by $\mathbf{P}$ then it changes the order of the rows of $\mathbf{A}$. If $\mathbf{A}$ is multiplied from the right by $\mathbf{P}$ then it changes the order of the columns of $\mathbf{A}$. A permutational matrix is a matrix that has only one non-zero element equal to one in each row and each column. The position of the non-zero element P_{ij} is such that the jth row must be placed in the ith position in $\mathbf{PA}$ and the ith column must be placed in the jth position in $\mathbf{AP}$. The permutation matrix has the property $\mathbf{P}^{-1} = \mathbf{P}^T$ and multiplication of two permutations is another permutation.

Example. If

$$\mathbf{P} = \begin{bmatrix} 0 & 1 & 0 & 0 \\ 0 & 0 & 1 & 0 \\ 1 & 0 & 0 & 0 \\ 0 & 0 & 0 & 1 \end{bmatrix} \text{ and } \mathbf{A} = \begin{bmatrix} A_{11} & A_{12} & A_{13} & A_{14} \\ A_{21} & A_{22} & A_{23} & A_{24} \\ A_{31} & A_{32} & A_{33} & A_{34} \\ A_{41} & A_{42} & A_{43} & A_{44} \end{bmatrix}$$

then

$$\mathbf{PA} = \begin{bmatrix} A_{21} & A_{22} & A_{23} & A_{24} \\ A_{31} & A_{32} & A_{33} & A_{34} \\ A_{11} & A_{12} & A_{13} & A_{14} \\ A_{41} & A_{42} & A_{43} & A_{44} \end{bmatrix} \text{ and } \mathbf{AP} = \begin{bmatrix} A_{13} & A_{11} & A_{12} & A_{14} \\ A_{23} & A_{21} & A_{22} & A_{24} \\ A_{33} & A_{31} & A_{32} & A_{34} \\ A_{43} & A_{41} & A_{42} & A_{44} \end{bmatrix}$$

4. A circulant matrix $(\mathbf{U})$ is a matrix having columns, $\{\mathbf{C}; \mathbf{P}_c\mathbf{C}; \mathbf{P}_c^2\mathbf{C}, \ldots, \mathbf{P}_c^{n-1}\mathbf{C}\}$ where $\mathbf{P}_c$ is a permutation matrix that performs cyclic permutation of the elements of vector $\mathbf{C}$.

$$\mathbf{P}_c = \begin{bmatrix} 0 & 0 & 0 & \cdots & 0 & 1 \\ 1 & 0 & 0 & \cdots & 0 & 0 \\ 0 & 1 & 0 & \cdots & 0 & 0 \\ \vdots & \vdots & \vdots & & \vdots & \vdots \\ 0 & 0 & 0 & \cdots & 1 & 0 \end{bmatrix}$$

Thus if $\{\mathbf{C}^T = C_1, C_2, \ldots, C_n\}$ then $(\mathbf{P}_c\mathbf{C})^T = \{C_n, C_1, C_2, \ldots, C_{n-1}\}$ and $(\mathbf{P}_c^2\mathbf{C})^T = \{C_{n-1}, C_n, C_1, \ldots, C_{n-2}\}$ and so on. The circulant matrix has the following form:

$$\mathbf{U} = \begin{bmatrix} C_1 & C_n & C_{n-1} & \cdots & C_2 \\ C_2 & C_1 & C_n & & C_3 \\ \vdots & \vdots & \vdots & & \vdots \\ C_n & C_{n-1} & C_{n-2} & \cdots & C_1 \end{bmatrix}$$

A circulant matrix is determined by its first column (or row). These matrices appear in several physical and chemical problems.

Appendix 2
A feedback shift register and the fast Hadamard transform

To construct a PRBS and also to develop the fast Hadamard transform (FHT) a feedback shift register is needed. In general, this is a connection of two stage elements and may be considered as consisting of p little boxes, representing computer memory elements or flip-flops, each containing a 0 or a 1. At each time unit the contents of the boxes are shifted one place to the right. Before the shift the content of a prespecified box, called a feedback element, is added to the content of the pth (right-most) box and fed, after the shift, into the first (left-most) box. The sum is calculated modulo 2 (i.e. $0+0=1+1=0$ and $1+0=0+1=1$).

The necessary feedback elements of the shift register are listed in Table A2.1.

Table A2.1 — Suitable feedback connections for generation of maximum length PRBS

Number of shift register elements	Period of sequence $n = 2^p - 1$	Feedback element
3	7	1
4	15	1
5	31	2
6	63	1
7	127	6
9	511	5
10	1023	7
11	2047	9
15	32767	1

The value of the right-most element of the register that was 'pulled out' during the shift is an element of the PRBS. This PRBS has the maximum possible length equal to $n = 2^p - 1$. It follows from Table A2.1, that simple feedback connections are not available for every p (e.g. $p = 8$ is absent). More extensive tables for feedback connections are given in [3,4].

App. 2]   A feedback shift register and the fast Hadamard transform   203

Let us generate the PRBS from a shift register with $p = 4$ elements. The corresponding shift register stages are given in Table A2.2. For convenience we take the initial loading of the shift register as 1, 1, 1, 1.

Table A2.2 — Generation of PRBS from 4-element shift register

State number	State 1	State 2	State 3	State 4	PRBS	Decimal equivalent of the shift register stage
1	1	1	1	1	1	15
2	0	1	1	1	1	7
3	1	0	1	1	1	11
4	0	1	0	1	1	5
5	1	0	1	0	0	10
6	1	1	0	1	1	13
7	0	1	1	0	0	6
8	0	0	1	1	1	3
9	1	0	0	1	1	9
10	0	1	0	0	0	4
11	0	0	1	0	0	2
12	0	0	0	1	1	1
13	1	0	0	0	0	8
14	1	1	0	0	0	12
15	1	1	1	0	0	14
16	1	1	1	1	1	15

Thus the 16th stage of the shift register is equal to the first one (i.e. the PRBS is periodic with $15 = 2^4 - 1$ elements). Considering the shift register stages as a set of binary numbers we can easily see that the shift register goes through all possible numbers from 1 to 15, though not in their natural order.

Consider now our input sequence matrix $\mathbf{X}$ in Eq. (2.4). Strictly speaking, that matrix is a right circulant matrix, i.e. all its rows are obtained from the first row by cyclically shifting its elements to the right and placing the last element in the position of the first element. A left circulant matrix is obtained similarly except that the cyclic shift is performed to the left and the first element is placed in the position of the last element. It is easy to verify that $\mathbf{M} = \mathbf{P}\mathbf{X}^T = \mathbf{X}\mathbf{P}$, where the index T means matrix transposition and $\mathbf{M}$ means left circulant matrix and $\mathbf{P}$ is a permutational matrix of the following form:

$$\mathbf{P} = \begin{bmatrix} 1 & 0 & \cdots & 0 & 0 \\ 0 & 0 & \cdots & 0 & 1 \\ 0 & 0 & \cdots & 1 & 0 \\ \cdot & \cdot & & \cdot & \cdot \\ \cdot & \cdot & & \cdot & \cdot \\ \cdot & \cdot & & \cdot & \cdot \\ 0 & 1 & & 0 & 0 \end{bmatrix}$$

Now we ask the reader to accept that a left circulant matrix can be factorized as

M = LS, where **L** is an $n \times p$ matrix obtained from the corresponding shift register stages (the contents of Table A2.2 are considered as a matrix), **S** is a $p \times n$ matrix, and matrix element computation in **M** must be done modulo 2. A rather complicated proof of this can be found in [5]. For $p = 3$ the corresponding factorization is:

$$\mathbf{M} = \begin{bmatrix} 1 & 1 & 1 & 0 & 1 & 0 & 0 \\ 1 & 1 & 0 & 1 & 0 & 0 & 1 \\ 1 & 0 & 1 & 0 & 0 & 1 & 1 \\ 0 & 1 & 0 & 0 & 1 & 1 & 1 \\ 1 & 0 & 0 & 1 & 1 & 1 & 0 \\ 0 & 0 & 1 & 1 & 1 & 0 & 1 \\ 0 & 1 & 1 & 1 & 0 & 1 & 0 \end{bmatrix} = \begin{bmatrix} 1 & 1 & 1 \\ 0 & 1 & 1 \\ 1 & 0 & 1 \\ 0 & 1 & 0 \\ 0 & 0 & 1 \\ 1 & 0 & 0 \\ 1 & 1 & 0 \end{bmatrix} \begin{bmatrix} 0 & 0 & 1 & 1 & 1 & 0 & 1 \\ 0 & 1 & 0 & 0 & 1 & 1 & 1 \\ 1 & 0 & 0 & 1 & 1 & 1 & 0 \end{bmatrix} = \mathbf{LS}$$

(A2.1)

Note also that:

$$\mathbf{XP} = \begin{bmatrix} 1 & 0 & 0 & 1 & 0 & 1 & 1 \\ 1 & 1 & 0 & 0 & 1 & 0 & 1 \\ 1 & 1 & 1 & 0 & 0 & 1 & 0 \\ 0 & 1 & 1 & 1 & 0 & 0 & 1 \\ 1 & 0 & 1 & 1 & 1 & 0 & 0 \\ 0 & 1 & 0 & 1 & 1 & 1 & 0 \\ 0 & 0 & 1 & 0 & 1 & 1 & 1 \end{bmatrix} \begin{bmatrix} 1 & 0 & 0 & 0 & 0 & 0 & 0 \\ 0 & 0 & 0 & 0 & 0 & 0 & 1 \\ 0 & 0 & 0 & 0 & 0 & 1 & 0 \\ 0 & 0 & 0 & 0 & 1 & 0 & 0 \\ 0 & 0 & 0 & 1 & 0 & 0 & 0 \\ 0 & 0 & 1 & 0 & 0 & 0 & 0 \\ 0 & 1 & 0 & 0 & 0 & 0 & 0 \end{bmatrix} = \begin{bmatrix} 1 & 1 & 1 & 0 & 1 & 0 & 0 \\ 1 & 1 & 0 & 1 & 0 & 0 & 1 \\ 1 & 0 & 1 & 0 & 0 & 1 & 1 \\ 0 & 1 & 0 & 0 & 1 & 1 & 1 \\ 1 & 0 & 0 & 1 & 1 & 1 & 0 \\ 0 & 0 & 1 & 1 & 1 & 0 & 1 \\ 0 & 1 & 1 & 1 & 0 & 1 & 0 \end{bmatrix} = \mathbf{M}$$

i.e. **M = XP** as claimed above. Also $\mathbf{P} = \mathbf{P}^T$.

The rows of the first multiplier, **L** in Eq. (A2.1) form the corresponding stages of the shift register, and the rows of the matrix **S** are formed from certain rows of the matrix **M**. The rows of **M** that form **S** are obtained by taking those rows of **M** that correspond to rows in **L** that contain only one non-zero element. The column containing the non-zero element in the row of **L** shows the row in **S** in which the row from **M** must be placed. For example, the fourth row of **L** contains one non-zero element, in the second column; we must choose the fourth row from **M** and put it into the second row in **S**. The next row in **L** with one non-zero element is the fifth and has it in the third column, so we must choose the fifth row of **M** and put it into the third row of **S**.

The factorization **M = LS** is important because it enables us to transform matrix **M** into a Hadamard matrix for which there is a fast multiplication algorithm. As was mentioned already, the rows of **L** (and also the columns of **S**) can be considered as binary numbers ranging from 1 to $n = 2^p - 1$. Let us add one row with zero elements

App. 2] A feedback shift register and the fast Hadamard transform

to **L** and one column with zero elements to **S**. Let $\mathbf{P}_1$ be a permutation matrix that arranges the matrix **L** rows in such a way that all corresponding binary numbers are ordered normally from 0 to n, and $\mathbf{P}_2$ is the matrix that arranges the columns of **S** in a similar way. Denote the matrix with its rows in ordered binary form by **B**. From Eq. (A2.1) we obtain $\mathbf{M} = \mathbf{P}_1^T \mathbf{B}\mathbf{B}^T \mathbf{P}_2^T$. The multiplication in $\mathbf{B}\mathbf{B}^T$ must also be computed by modulo 2. Denote it by $\mathcal{H}$. For $p = 3$, $\mathbf{B}\mathbf{B}^T$ has the following form:

$$\mathcal{H} = \begin{bmatrix} 0 & 0 & 0 \\ 0 & 0 & 1 \\ 0 & 1 & 0 \\ 0 & 1 & 1 \\ 1 & 0 & 0 \\ 1 & 0 & 1 \\ 1 & 1 & 0 \\ 1 & 1 & 1 \end{bmatrix} \begin{bmatrix} 0 & 0 & 0 & 1 & 1 & 1 & 1 \\ 0 & 0 & 1 & 1 & 0 & 0 & 1 & 1 \\ 0 & 1 & 0 & 1 & 0 & 1 & 0 & 1 \end{bmatrix} = \begin{bmatrix} 0 & 0 & 0 & 0 & 0 & 0 & 0 & 0 \\ 0 & 1 & 0 & 1 & 0 & 1 & 0 & 1 \\ 0 & 0 & 1 & 1 & 0 & 0 & 1 & 1 \\ 0 & 1 & 1 & 0 & 0 & 1 & 1 & 0 \\ 0 & 0 & 0 & 0 & 1 & 1 & 1 & 1 \\ 0 & 1 & 0 & 1 & 0 & 1 & 0 & 1 \\ 0 & 0 & 1 & 1 & 1 & 1 & 0 & 0 \\ 0 & 1 & 1 & 0 & 1 & 0 & 0 & 1 \end{bmatrix}$$

Replacing in $\mathcal{H}$ all zeros by ones and all ones by -1 we get a Hadamard matrix, $\widetilde{\mathcal{H}}$, of order p. It is easy to see that $\widetilde{\mathcal{H}} = -2\mathcal{H} + \mathbf{J}$. By definition the Hadamard matrix can be constructed by recurrence:

$$\widetilde{\mathcal{H}}(0) = 1; \quad \widetilde{\mathcal{H}}(p) = \begin{bmatrix} \widetilde{\mathcal{H}}(p-1) & \widetilde{\mathcal{H}}(p-1) \\ \widetilde{\mathcal{H}}(p-1) & -\widetilde{\mathcal{H}}(p-1) \end{bmatrix}$$

Thus

$$\widetilde{\mathcal{H}}(1) = \begin{bmatrix} 1 & 1 \\ 1 & -1 \end{bmatrix} \text{ and } \widetilde{\mathcal{H}}(2) = \begin{bmatrix} 1 & 1 & | & 1 & 1 \\ 1 & -1 & | & 1 & -1 \\ - & - & & - & - \\ 1 & 1 & | & -1 & -1 \\ 1 & -1 & | & -1 & 1 \end{bmatrix}$$

The dashed lines in $\widetilde{\mathcal{H}}(2)$ bound the $\widetilde{\mathcal{H}}(1)$ matrices. $\widetilde{\mathcal{H}}(2)$ was used already in the weighing experiment in Section 2.1.

The matrices are named after French mathematician J. Hadamard who was the first to study them. The Hadamard matrix has the property $\widetilde{\mathcal{H}}^T\widetilde{\mathcal{H}} = (n+1)\mathbf{I}$ and the multiplication of a matrix **A** by a Hadamard matrix can be done by a fast algorithm, the structure of which is very similar to the fast Fourier transform (FFT) except that in the FHT only additions and subtractions are necessary and thus it is much faster than the FFT. The algorithms can be found in [6,7]. A program listing is given in

Appendix 3. Usually multiplication of a vector with a matrix requires $(n+1)^2$ operations. The fast algorithm requires $(n+1)\log_2(n+1)$ arithmetic operations. For example, if $n = 1023$ as is common, then in CC the decorrelation time can be reduced by approximately two orders of magnitude, which is a very significant time saving.

The decorrelation algorithm can now be derived. A Hadamard matrix is a matrix of order 2^p when the **X** matrix in Eq. (2.4) is of order $n = 2^p - 1$. Thus one artificial row and column should be added to the **X** matrix to transform it into a Hadamard matrix. Let us take the elements of this row and column as all equal to zero. Also one artificial element $Y_0 = 0$ must be added to the output vector and one artificial element H_0 with arbitrary value must be added to the chromatogram vector. Denote the extended input matrix by $\tilde{\mathbf{X}}$ and the extended output and chromatogram vectors by $\tilde{\mathbf{Y}}$ and $\tilde{\mathbf{H}}$ respectively. The output vector **Y** can now be transformed as follows:

(1) $\quad \tilde{\mathbf{Y}} = \tilde{\mathbf{X}}\tilde{\mathbf{H}} = \tilde{\mathbf{X}}\mathbf{P}\,\mathbf{P}\tilde{\mathbf{H}} = \mathbf{M}(\mathbf{P}\tilde{\mathbf{H}})$
$\quad\quad = \mathbf{LS}(\mathbf{P}\tilde{\mathbf{H}}) = (\mathbf{P}_1^T \mathbf{B})(\mathbf{B}^T \mathbf{P}_2^T)(\mathbf{P}\tilde{\mathbf{H}})$
$\quad\quad = \mathbf{P}_1^T \mathbf{B}\mathbf{B}^T \mathbf{P}_2^T (\mathbf{P}\tilde{\mathbf{H}}) = \mathbf{P}_1^T \, \mathcal{H} \, \mathbf{P}_2^T (\mathbf{P}\tilde{\mathbf{H}})$

(2) The FHT is applied to the permuted output sequence:

$$\mathcal{H}^T \mathbf{P}_1 \tilde{\mathbf{Y}} = -\frac{n+1}{2} \mathbf{P}_2^T (\mathbf{P}\tilde{\mathbf{H}}) + \frac{1}{2} \mathcal{H} \, \mathbf{J} \, \mathbf{P}_2^T (\mathbf{P}\tilde{\mathbf{H}})$$

The product $\mathcal{H}\mathbf{J}$ is a matrix having only the zeroth row with non-zero elements, each equal to $n+1$, and $[\tilde{\mathbf{H}} \, \mathbf{J} \, \mathbf{P}_2^T (\mathbf{P}\tilde{\mathbf{H}})]/2$ is a vector having only one non-zero element in the zeroth position, equal to $(n+1)(\Sigma H_i)/2$. Hence it influences only the zeroth element of the extended chromatogram vector. That element does not contain analytical information and may be neglected.

(3) The permuted and transformed output vector is permuted with $\mathbf{PP}_2$ and normalized with the factor $(n+1)/2$.

$$\tilde{\mathbf{H}} = -\frac{2}{n+1} \mathbf{PP}_2 \, \mathcal{H} \, \mathbf{P}_1 \, \tilde{\mathbf{Y}} \tag{A2.2}$$

Equation (A2.2) is the desired decorrelation algorithm required by the FHT. Since $\mathbf{M} = \mathbf{M}^T$ and $\tilde{\mathbf{H}} = -\frac{2}{n+1} \mathbf{PP}_1^T \, \mathcal{H} \mathbf{P}_2^T \, \tilde{\mathbf{Y}}$ then that provides the other possibility for computing a chromatogram. Permutations **P**, $\mathbf{P}_1$ and $\mathbf{P}_2$ can be generated beforehand according to the rules described above, stored on a disc as a sequence of numbers, and used every time that decorrelation is necessary. Multiplication of a vector by a permutation matrix is simply a rearrangement of the vector elements. The FHT can be performed by standard algorithms. Thus, although having a complicated theoretical background the algorithm actually requires only the rearrangement of the output sequence elements, transformation, and then rearrangement of the chromatogram vector elements to get the desired correlogram. The algorithm described above was first proposed in Hadamard transform optical and NMR spectrometry [8, 9].

App. 2] A feedback shift register and the fast Hadamard transform

Because FHT uses only subtractions and additions, the algorithm can be programmed quite easily in computer machine code. This ensures that decorrelation by FHT is almost instantaneous on an Apple II microcomputer even for $(n+1) = 512$–2048.

Let us consider an example of decorrelation by fast Hadamard transform. Let the chromatogram vector be $\mathbf{H}^T = (0,0,1,0,0,2,0)$.

The input sequence with 7 elements in $\mathbf{X}^T = (1,1,1,0,1,0,0)$. From Eq. (A2.2) we get the output sequence as follows (beginning from Y_8): $\mathbf{Y}^T = (2,0,3,1,1,2,3)$. The extended matrix equation $\widetilde{\mathbf{Y}} = \widetilde{\mathbf{X}}\widetilde{\mathbf{H}}$ is:

$$\begin{bmatrix} 0 \\ 2 \\ 0 \\ 3 \\ 1 \\ 1 \\ 2 \\ 3 \end{bmatrix} = \begin{bmatrix} 0 & 0 & 0 & 0 & 0 & 0 & 0 \\ 0 & 1 & 0 & 0 & 1 & 0 & 1 & 1 \\ 0 & 1 & 1 & 0 & 0 & 1 & 0 & 1 \\ 0 & 1 & 1 & 1 & 0 & 0 & 1 & 0 \\ 0 & 0 & 1 & 1 & 1 & 0 & 0 & 1 \\ 0 & 1 & 0 & 1 & 1 & 1 & 0 & 0 \\ 0 & 0 & 1 & 0 & 1 & 1 & 1 & 0 \\ 0 & 0 & 0 & 1 & 0 & 1 & 1 & 1 \end{bmatrix} \begin{bmatrix} H_0 \\ 0 \\ 0 \\ 1 \\ 0 \\ 0 \\ 2 \\ 0 \end{bmatrix}$$

This equation can be transformed to the following:

$$\widetilde{\mathbf{H}} = \frac{2}{n+1} \mathbf{P} \begin{bmatrix} 1 & 0 & 0 & 0 & 0 & 0 & 0 & 0 \\ 0 & 1 & 0 & 0 & 0 & 0 & 0 & 0 \\ 0 & 0 & 1 & 0 & 0 & 0 & 0 & 0 \\ 0 & 0 & 0 & 0 & 1 & 0 & 0 & 0 \\ 0 & 0 & 0 & 0 & 0 & 1 & 0 & 0 \\ 0 & 0 & 0 & 0 & 0 & 0 & 0 & 1 \\ 0 & 0 & 0 & 1 & 0 & 0 & 0 & 0 \\ 0 & 0 & 0 & 0 & 0 & 0 & 1 & 0 \end{bmatrix} \widetilde{\mathcal{H}} \begin{bmatrix} 1 & 0 & 0 & 0 & 0 & 0 & 0 & 0 \\ 0 & 0 & 0 & 0 & 0 & 1 & 0 & 0 \\ 0 & 0 & 0 & 0 & 1 & 0 & 0 & 0 \\ 0 & 0 & 1 & 0 & 0 & 0 & 0 & 0 \\ 0 & 0 & 0 & 0 & 0 & 0 & 1 & 0 \\ 0 & 0 & 0 & 1 & 0 & 0 & 0 & 0 \\ 0 & 0 & 0 & 0 & 0 & 0 & 0 & 1 \\ 0 & 1 & 0 & 0 & 0 & 0 & 0 & 0 \end{bmatrix} \widetilde{\mathbf{Y}}$$

Having permutations $\mathbf{P}, \mathbf{P}_1$ and $\mathbf{P}_2$, Eq. (A2.2) can be applied directly as follows:

(1) $(\mathbf{P}_1 \widetilde{\mathbf{Y}})^T$ = $\{0, 1, 1, 0, 2, 3, 3, 2\}$
(2) $(\widetilde{\mathcal{H}} \mathbf{P}_1 \widetilde{\mathbf{Y}})^T$ = $\{12, 0, 0, -4, -8, 0, 0, 0\}$
(3) $(\mathbf{P}_2 \widetilde{\mathcal{H}} \mathbf{P}_1 \widetilde{\mathbf{Y}})^T$ = $\{12, 0, 0, -8, 0, 0, -4, 0\}$
(4) $(\mathbf{P}\mathbf{P}_2 \widetilde{\mathcal{H}} \mathbf{P}_1 \widetilde{\mathbf{Y}})^T$ = $\{12, 0, 0, -4, 0, 0, -8, 0\}$
(5) $-2(\mathbf{P}\mathbf{P}_2 \widetilde{\mathcal{H}} \mathbf{P}_1 \widetilde{\mathbf{Y}})^T/(n+1)$ = $\{3, 0, 0, 1, 0, 0, 2, 0\}$

Thus it follows from this algorithm test that H_0 takes the value ΣH_i. Steps (3) and (4) are in practice joined together and the permutation $\mathbf{PP}_2$ is stored on disc or generated by an algorithm. Here we separate their function for clarity. Normalization with the factor $-2/(n+1)$ is not usually necessary if only relative chromatographic peak intensities are of interest.

Appendix 3
Software for the correlation chromatography

Appendix 3.1   Generation of permutations

First we describe an Applesoft Basic program for the generation of the permutations **PP**$_1$ and **P**$_2$. The program listing is given in Fig. A3.1 and the sample run for $p = 3–5$ in Fig. A3.2.

The program variables have the following meaning:

KST(I)	— shift register content
KQ(I)	— feedback connections
KIT(I)	— rows in matrix **L** that contain only one non-zero element
PER$(I)	— permutation file name
J1%(I)	— **PP**$_1$ permutation
J2%(I)	— **P**$_2$ permutation
MX%(I)	— PRBS vector
NASTE	— the number of boxes in the sh ft register. $N = 2^{NASTE}$

Statement 110 gives the look-up table for feedback connections.

Statements 200–360 generate the PRBS and **P**$_1$ permutation as a set of decimal numbers computed from their binary equivalents in the shift register. The index I in the J1%(I) array gives the row number in the **P**$_1$ matrix and the non-zero element column number in this row is expressed by the value of J1%(I).

Statement 230 transforms the binary content of the shift register into a decimal number.

Statement 270 finds the row numbers in the **L** matrix that contain only one non-zero element, and these numbers are used for construction of the **P**$_2$ matrix.

Statements 370–460 generate the **P**$_2$ permutation. The sequence J2%(I) elements are interpreted in the same way as those in J1%(I).

Statements 470–510	find the permutation **PP**$_1$.
Statements 520–550	perform the output to printer.
Statements 560–620	perform the output to disc according to Apple Dos 3.3 requirements.

```
UPR#0
ULOAD PERMUGEN1
ULIST
100  DIM KIT(12),KST(12),KQ(12),PER$(12)
105  DIM J1%(1025),J2%(1025),MX%(1025)
110  DATA    1,1,1,1,3,5,6,0,5,7,9
120  FOR I = 1 TO 11: READ KQ(I): NEXT I
122  DATA  "P1","P2","P3","P4","P5","P6","P7","P8","P9","P10","P11"
124  FOR K = 1 TO 11: READ PER$(K): NEXT K
130  INPUT "SEQUENCE ";NASTE
137  IF NASTE = 8 THEN  GOTO 900
140  N = 2 ^ NASTE
145  KREG = KQ(NASTE)
150  NN = N - 1
160  NISTE = NASTE - 1
170  FOR K = 1 TO NASTE:KST(K) = 1: NEXT K
180  REM
190  REM
200  FOR I = 1 TO NN
210  J1%(I) = 0
220  MX%(I) = KST(NASTE)
230  FOR K = 1 TO NASTE:J1%(I) = J1%(I) + 2 ^ (NASTE - K) * KST(K): NEXT K
260  FOR J = 1 TO NASTE
270  IF 2 ^ (J - 1) - J1%(I) = 0 THEN KIT(J) = I
280  NEXT J
290  ISUM = KST(KREG) + KST(NASTE)
300  IF ISUM = 1 THEN  GOTO 320
310  ISUM = 0
320  FOR J = 1 TO NISTE:KST(NASTE + 1 - J) = KST(NASTE - J): NEXT J
350  KST(1) = ISUM
360  NEXT I
362  REM
364  REM
370  FOR I = 1 TO NN
380  J2%(I) = 0
390  FOR K = 1 TO NASTE
400  KK = KIT(K)
410  J2%(I) = J2%(I) + MX%(KK) * 2 ^ (K - 1)
420  NEXT K
430  MVAH = MX%(1)
440  FOR J = 2 TO NN:MX%(J - 1) = MX%(J): NEXT J
450  MX%(NN) = MVAH
460  NEXT I
462  REM
464  REM
470  IP = N / 2
480  FOR I = 2 TO IP
490  IVA = J1%(I)
500  J1%(I) = J1%(N + 1 - I):J1%(N + 1 - I) = IVA
510  NEXT I
520  PRINT "SEQUENCE=";NASTE
522  PRINT
525  PRINT "NO";"   ";"PRBS";
526  PRINT "     ";"PP1";"     ";"P2"
527  PRINT : PRINT : PRINT
530  FOR U = 1 TO NN
540  PRINT U;"     ";;MX%(U);
542  PRINT "        ";J1%(U);"      ";J2%(U)
550  NEXT U
555  PRINT
556  PRINT
560  D$ =  CHR$ (4)
570  PRINT D$;"OPEN";PER$(NASTE)
580  PRINT D$;"WRITE";PER$(NASTE)
590  FOR J = 1 TO NN
600  PRINT J1%(J)
602  PRINT J2%(J)
610  NEXT J
620  PRINT D$;"CLOSE";PER$(NASTE)
900  END
```

Fig. A3.1 — Permutation generation program.

SEQUENCE=3

NO	PRBS	PP1	P2
1	1	7	1
2	1	6	2
3	1	4	4
4	0	1	5
5	1	2	7
6	0	5	3
7	0	3	6

SEQUENCE=4

NO	PRBS	PP1	P2
1	1	15	1
2	1	14	2
3	1	12	4
4	1	8	8
5	0	1	9
6	1	2	11
7	0	4	15
8	1	9	7
9	1	3	14
10	0	6	5
11	0	13	10
12	1	10	13
13	0	5	3
14	0	11	6
15	0	7	12

SEQUENCE=5

NO	PRBS	PP1	P2
1	1	31	1
2	1	30	2
3	1	28	4
4	1	25	8
5	1	19	16
6	0	6	5
7	0	13	10
8	0	26	20
9	1	20	13
10	1	9	26
11	0	18	17
12	1	4	7
13	1	8	14
14	1	16	28
15	0	1	29
16	1	2	31
17	0	5	27
18	1	10	19
19	0	21	3
20	0	11	6
21	0	23	12
22	0	14	24
23	1	29	21
24	0	27	15
25	0	22	30
26	1	12	25
27	0	24	23
28	1	17	11
29	1	3	22
30	0	7	9
31	0	15	18

Fig. A3.2 — Some permutations.

Appendix 3.2 Experiment control program

This program performs the PRBS input to the chromatograph and acquires the detector output data as a sequence of output vector elements. Because this program usually includes input devices controlling drivers and ADC subroutines that are different for different equipment, a simulation program is given that generates the output vector by means of software. The listing is given in Fig. A3.3.

The program variables have the following meanings:

CR%(I)	— chromatogram vector
A%(I)	— peak height
RT%(I)	— retention time
S%(I)	— peak width
OP%(I)	— detector output
KR(I)	— shift register
NASTE	— sequence order ($N = 2^{NASTE}$)
NP	— number of peaks
S1,S2	— the beginning and end of a peak

Statements 100–300 generate a single injection chromatogram.

Statement 730 gives the feedback elements.

Statement 740 is the initialization of the feedback shift register.

Statements 800–920 are the generation of two PRBSs and the detector outputs. The second output sequence is written on disc.

Statement 820 checks the running PRBS value. If this value is 1 then the chromatogram is added to the detector output in statements 826–840.

Statements 860–900 generate the following PRBS element.

Statements 2140–2310 perform the output to disc according to Apple Dos 3.3 requirements.

Appendix 3.3 Decorrelation program

The program listing is given in Fig. A3.4. This program decorrelates the output vector by fast Hadamard transform. The meanings of the variables are as follows:

Y(I)	— the output vector and correlogram
RY(I)	— input vector to FHT
J1%(I), J2%(I), PER$(I),NASTE	— determined already in A 3.1.
A(I)	— a display plot of the correlogram
CD.DAT	— the output vector elements file name
BC	— display variable

Statements 5–16 are integer variable declarations according to the conventions of the Microsoft TASC compiler.

Statements 50–82 perform the fast Hadamard transform subroutine according to Kunt [7].

Statements 120–290 input the **Y** vector elements and permutations PP_1^T and P_2^T from disc files according to the conventions of Apple Dos 3.3.

Statements 630–650 make the permutation of **Y** by P_2^T.

Statement 660 — $Y_0 = 0$.

Statement 670 is FHT.

Statements 680–700 make permutation of $\mathcal{H}\mathbf{P}_2^T\mathbf{Y}$ by $\mathbf{PP}_1^T$
Statements 812–818 normalize the correlogram by $-2/(n+1)$
Statements 920–1072 perform drawing of the correlogram in the Apple IIe display.

```
]LOAD CORREDEMO1
]LIST

5    REM   CC SIMULAATOR
8    REM   WRITTEN FOR THE BOOK
10   DIM CR%(279),A%(15),RT%(15),S%(15)
15   DIM OP%(2050),KR(11)
17   PRINT "THIS PROGRAMM SIMULATES THE WORK OF"
18   PRINT "THE CORRELATION CHROMATOGRAPHY"
19   PRINT
20   PRINT "WRITTEN BY MIHKEL KALJURAND IN MAY 1982"
21   PRINT
100  PRINT "LET US GENERATE A CHROMATOGRAMM"
101  PRINT
120  INPUT "NR. OF PEAKS (N<15) ";NP
130  PRINT "GIVE PEAK HEIGHT,RET.TIME AND WIDTH"
132  PRINT "RT<=60 RECOMENDED"
140  FOR I = 1 TO NP
150  PRINT "PEAK ";I
160  INPUT A%(I),RT%(I),S%(I)
170  NEXT I
200  FOR I = 1 TO NP
210  S1 = RT%(I) - 3 * S%(I)
220  S2 = RT%(I) + 3 * S%(I)
230  IF S1 < 0 THEN S1 = 0
240  IF S2 > 279 THEN S2 = 279
250  FOR K = S1 TO S2
260  U = (K - RT%(I)) / (1.414 * S%(I))
270  CR%(K) = CR%(K) + A%(I) * EXP ( - (U ^ 2))
280  NEXT K
290  NEXT I
300  PRINT
308  INPUT "THE SEQUENCE IS ";NASTE: REM  INPUT LOG2(N)
710  PRINT
730  DATA    0,1,1,1,2,1,6,0,5,7,9
740  FOR I = 1 TO 11: READ KR(I): NEXT I
750  TBS = KR(NASTE)
760  N = 2 ^ NASTE - 1
762  N2 = N * 2
770  FOR I = 1 TO NASTE:KR(I) = 1: NEXT I
800  FOR J = 0 TO 1
810  FOR I = 1 TO N
820  IF KR(NASTE) = 0 THEN  GOTO 860
826  FOR L = 1 TO S2
828  IF J * N + I + L < N THEN  GOTO 840
829  IF J * N + I + L > N2 THEN  GOTO 860
832  OP%(J * N + I + L) = OP%(J * N + I + L) + CR%(L)
840  NEXT L
860  IF KR(NASTE) < > KR(TBS) THEN FB = 1
870  IF KR(NASTE) = KR(TBS) THEN FB = 0
880  FOR M = NASTE - 1 TO 1 STEP  - 1
890  KR(M + 1) = KR(M)
900  NEXT M
905  KR(1) = FB
907  PRINT J;" ";I;"     ";KR(NASTE),OP%(J * N + I)
910  NEXT I
920  NEXT J
2140 D$ =  CHR$ (4)
2150 PRINT D$;"OPEN  CD.DAT"
2160 PRINT D$;"WRITE CD.DAT"
2162 PRINT S2
2170 FOR I = 1 TO S2
2172 PRINT CR%(I)
2174 NEXT I
2188 PRINT NASTE
2190 FOR I = N + 1 TO N2
2192 PRINT OP%(I)
2194 NEXT I
2300 PRINT D$;"CLOSE CD.DAT"
```

Fig. A3.3 -- Experiment control program.

```
ÜLOAD DECORDEMO1
ÜLIST

1    REM   DECORRELATION DEMO
5    REM ! INTEGER J,R,I,NASTE,IA,IB,K,N,IC,NN,N2,N4,BL
7    REM ! INTEGER Y(2048),CR%(100)
16   REM ! INTEGER J1%(1024),J2%(1024)
30   GOTO 111
40   REM
50   REM   FHT
52   FOR I = 1 TO NASTE
53   PRINT I
54   IA = 2 ^ I
56   IB = IA / 2
58   FOR J = 1 TO IB
60   FOR K = J TO N STEP IA
62   IC = K + IB
64   IZ = RY(IC)
66   RY(IC) = RY(K) - IZ
68   RY(K) = RY(K) + IZ
70   NEXT K
72   NEXT J
80   NEXT I
82   RETURN
111  DIM PER$(12),RY(1024)
120  D$ = CHR$ (4)
122  PRINT D$;"OPEN  CD.DAT"
124  PRINT D$;"READ  CD.DAT"
126  INPUT S2
127  FOR I = 2 TO S2 + 1
128  INPUT CR%(I)
129  NEXT I
136  INPUT NASTE
137  N = 2 ^ NASTE:NN = N - 1
138  FOR I = 1 TO NN
139  INPUT Y(I)
140  NEXT I
142  PRINT D$;"CLOSE CD.DAT"
210  DATA   "P1","P2","P3","P4","P5","P6","P7","P8","P9","P10"
220  FOR K = 1 TO 10: READ PER$(K): NEXT K
228  D$ =  CHR$ (4)
230  PRINT D$;"OPEN";PER$(NASTE)
240  PRINT D$;"READ";PER$(NASTE)
250  FOR I = 1 TO NN
260  INPUT J1%(I)
270  INPUT J2%(I)
280  NEXT I
290  PRINT D$;"CLOSE";PER$(NASTE)
630  FOR I = 1 TO NN
640  RY(J2%(I) + 1) = Y(I)
650  NEXT I
660  RY(1) = 0
670  GOSUB 50: REM   FHT
680  FOR I = 1 TO NN
690  Y(I) = RY(J1%(I) + 1)
700  NEXT I
710  REM
812  FOR I = 1 TO NN
814  Y(I) = - (Y(I) / (NN + 1)) * 2
816  PRINT I,Y(I)
818  NEXT I
1040 HGR
1045 HPLOT 0,0
1050 FOR I = 0 TO 255
1070 HPLOT  TO I,Y(I)
1072 NEXT I
```

Fig. A3.4 — Decorrelation program.

The program described here can be compiled (e.g. by using the Microsoft TASC compiler) and then executed as a machine code program. This speeds up decorrelation remarkably. However, care must be taken not to mix the memory areas for graphics, program, and graphic variables.

Appendix 4
Contour plot BASIC subroutine

The program listing is given in Fig. A4.1.

Definition of variables:
- $Z\%(I,J)$ — two-dimensional function values at the corner points of the unit square
- CA — cut level (cutting plane + co-ordinate)
- IE — running index of the second variable (e.g. chromatogram number)
- UB — running index of the first variable (e.g., particular point of IEth chromatogram)
- XO, YO — co-ordinates of the 0,0 corner of the unit square
- DX, DY — increments of the first and second variables
- X, Y — co-ordinates of the cutting line between surface and plane
- P\$ — plotter pen control variable: P\$='U' means 'pen up' and P\$='D' means 'pen down'.

Statements 630–710 compare the cutting level with the actual value of the function over a particular unit square. If the cutting level is higher or lower than the surface, the program skips this unit square.

Statements 790–1020 compute the cutting line co-ordinates.

Statements 855–875 and 980–1010 are dependent on plotter type and should be replaced by the user.

In use of this subroutine the variables IE and UB are run through all possible values minus one. GO SUB 300 is a plotter driver subroutine.

```
ûLIST

578 REM  CONTOUR SUBROUTINE
590 Y0 = DY * IE
610 X0 = DX * UB
620 NC = 0
630 Z1 = Z%(0,0)
640  IF CA > Z1 THEN NC = NC + 1
650 Z2 = Z%(0,1)
660  IF CA > Z2 THEN NC = NC + 1
670 Z3 = Z%(1,0)
680  IF CA > Z3 THEN NC = NC + 1
690 Z4 = Z%(1,1)
700  IF CA > Z4 THEN NC = NC + 1
710  IF NC = 0 OR NC = 4 THEN  GOTO 1040
720 A = Z1:B = Z2 - A:C = Z3 - A
750 D = Z4 - A - B - C
760 KT = 0
790  FOR X1 = 0 TO 1 STEP 0.25
800 DS = C + D * X1
810  IF DS = 0 THEN  GOTO 880
820 Y1 = (CA - A - B * X1) / DS
830  IF Y1 < 0 OR Y1 > 1 THEN  GOTO 880
840 X = X0 + DX * X1
850 Y = Y0 + DY * Y1
855  IF X > 9999 OR Y > 9999 THEN  GOTO 875
860  IF KT = 0 THEN P$ = "U": GOSUB 300
870 P$ = "D": GOSUB 300
875 KT = KT + 1
880  NEXT X1
884 P$ = "U": GOSUB 300
900 KT = 0
910  FOR Y1 = 0 TO 1 STEP 0.25
920 DR = B + D * Y1
930  IF DR = 0 THEN  GOTO 1020
940 X1 = (CA - A - C * Y1) / DR
950  IF X1 < 0 OR X1 > 1 THEN  GOTO 1020
960 X = X0 + DX * X1
970 Y = Y0 + DY * Y1
980  IF X > 9999 OR Y > 9999 THEN  GOTO 1010
990  IF KT = 0 THEN P$ = "U": GOSUB 300
1000 P$ = "D": GOSUB 300
1010 KT = KT + 1
1020  NEXT Y1
1022 P$ = "U": GOSUB 300
1040  RETURN
```

Fig. A4.1

APPENDIX REFERENCES

[1] G. Strang, *Linear Algebra and its Applications*, Academic Press, New York, 1976.
[2] R. Bellman, *Introduction to Matrix Analysis*, McGraw-Hill, New York, 1960.
[3] G. A. Korn, *Random-Process Simulation and Measurements*, McGraw-Hill, New York, 1966.
[4] W. W. Peterson, *Error Correcting Codes*, MTI Press, Cambridge, MA, 1961.
[5] M. Cohn and A. Lempel, *IEEE Trans. Inf. Theory*, 1977, **IF-23**, 135.
[6] N. Ahmed, K. R. Rao and M. Abdussattar, *IEEE Trans. Audio Electroacoustics*, 1971, **AU-19**, 225.
[7] M. Kunt, *IEEE Trans. Computers*, 1975, **C-24**, 1120.
[8] E. D. Nelson and M. L. Fredman, *J. Opt. Soc. Am.*, 1970, **60**, 1664.
[9] A. Kaiser, *J. Mag. Reson.*, 1974, **15**, 44.

Notation

Unfortunately, several notations used in traditional nomenclature have more than one meaning. We have not tried to remove this ambiguity because we do not think it should lead to any confusion, since no symbols with a double meaning are used in the same equation or even in the same paragraph.

A	— area, Golay equation constant
$\mathbf{A}$	— a matrix
$\mathbf{A}^T$	— transpose of a matrix
A^{-1}	— inverse of a matrix
B	— Golay equation constant
b	— rate of a linear process (baseline drift or input change), modulator length
$C_{x,y}(t)$	— correlation function
$c(t)$	— concentration
C_0	— initial concentration
C	— Golay equation constant
D	— diffusion coefficient, Golay equation constant
d	— diameter
E	— activation energy
$\mathbf{E}$	— input sequence error matrix
E_i	— input sequence element error
F	— volumetric flow-rate
$\mathbf{F}$	— Fourier transform matrix
f	— frequency
f_i	— response factor of a compound
$g(t)$	— amount of gas evolved at time t
g_0	— total amount of evolved gas
$\mathbf{H}$	— chromatogram vector
$\mathbf{H}$	— chromatogram matrix
H	— entropy, height equivalent to a theoretical plate
H_j	— chromatogram vector element
$H_j(i)$	— 2D chromatogram element

Notation

$\Delta \boldsymbol{H}$	— correlation noise vector
ΔH_i	— correlation noise vector element
$\overline{\Delta H}$	— mean value of ΔH element
$\mathcal{H}$	— Hadamard transform matrix
$h(t)$	— chromatogram
i, j	— integer index
$\mathbf{I}$	— unity (identity) matrix
I	— amount of information
$\mathbf{J}$	— matrix with all elements equal to 1
k	— capacity factor
$\mathbf{L}$	— matrix factor
l	— diffusion path length
M	— molecular weight
$\mathbf{M}$	— circulant matrix
m	— mass
N	— number of theoretical plates
n	— number of chromatogram elements
P	— total pressure in diffusion cell
$\mathbf{P}$	— permutation matrix
p	— pressure, probability
p_i	— initial pressure
p_a	— midpoint pressure in tandem-column systems
p_0	— outlet pressure
Q	— peak capacity of a chromatogram
q	— least significant bit value in ADC; an integer
R	— sample radius, gas constant, peak resolution
$\boldsymbol{R}$	— detector signal disturbance vector
$\bar{R}$	— mean value of the vector $\boldsymbol{R}$ elements
r	— diffusion rate, radius
$r(t)$	— random function
$\mathbf{S}$	— matrix factor
$s_{\Delta H}$	— correlation noise standard deviation
T	— temperature
t	— time
t_R	— retention time
t'	— adjusted retention time
t_m	— gas hold-up time
Δt	— time interval (sampling time, sampling interval, digitization time)
$\mathbf{U}$	— circulant matrix with PRBS elements
u	— linear gas velocity
V	— volume
V_j	— comparison vector element for removing hidden lines
$w_{1/2}$	— peak half-width
$x(t)$	— input function
$\mathbf{X}$	— input sequence matrix
X_i	— matrix $\mathbf{X}$ element
$y(t)$	— detector output signal

Y	— detector output vector
Y_i	— vector Y element
Z	— pre-exponential factor in Arrhenius equation
z	— spike height
γ	— a dimensionless constant
$\delta(t)$	— delta function
ε	— viscosity
ζ	— density
σ	— noise or peak standard deviation
τ	— time

ABBREVIATIONS

ADC	— analogue-to-digital converter
CC	— correlation chromatography
DAC	— digital-to-analogue converter
FT{ }	— Fourier transform operation
FT	— Fourier transform
FFT	— fast Fourier transform
FHT	— fast Hadamard transform
GC	— gas chromatography
HT	— Hadamard transform; heating time
IR	— infrared (spectrometry)
HETP	— height equivalent to a theoretical plate
HPLC	— high-pressure liquid chromatography
LC	— liquid chromatography
PRBS	— pseudo random binary sequence
PyGC	— pyrolysis gas chromatography
MS	— mass spectrometry
TG	— thermogravimetry
ThGC	— thermochromatography
TP	— temperature programming
TRT	— temperature rise-time

Index

abbreviations, 219
acrylonitrile–butadiene copolymer, 167
acrylonitrile rubber, 166
activation energy, 108
additive noise, 23, 42
adsorption in column, elimination, 18
adsorption sample concentration, 55
adsorption sites, 192
alcohols in water, 192
aliasing, 62
amount of information, 10, 84ff
   in two-dimensional systems, 90–92
analogue-to-digital conversion, 47, 61
analogue-to-digital converter resolution, 136, 137
aniline, 122
anticonvulsant drugs, 79
Apple computer, 94, 171
Applesoft BASIC, 97, 171
Arrhenius equation, 108
autocorrelation, 23–25, 39
automatic control, 9, 15, 187
   sampling, 118
autosampler, 81, 133

backflush, 76, 126
band-limited signal, 62
baseline drift, 42, 47, 52
   oscillations, 47
bilinear surface, 96, 97
biological molecules, 153
bit, 87
black box, 32
blood plasma, 79
boxcar chromatography, 73, 134,

capacity factor, 67, 93
capillary column, 136
carrier gas flow, 29
   velocity, 67
catalytic reactions, 72
cellulose, 181
chemical microreactor, 104
chemometrics, 9, 13, 157
chlorinated aromatics, 79

dibenzo-p-dioxins, 76
hydrocarbons, 86
phenols, 193, 194
polyethylene, 183
chloroprene, 166, 167, 177
chlorosulphonated polyethylene, 183
chromatogram, matrix, 68–70
   noise vector, 42
   resolution, 54, 55
   stopped-flow, 20
   two-dimensional, 59, 68, 72, 94
   vector, 32, 38
chromatograph, column, 15
   continuous, 21
   demands of multiple-input, 134–136
   detector, 15
   injector, 15
chromatography, boxcar, 73, 134
   capillary, 65, 67
   computerized multiple-input, 100, 137, 139, 170
   correlation, 28, 34, 37, 68, 131, 136, 163, 165, 166, 189, 192, 193
   differential, 19, 22, 190
   frequency modulation, 20–22
   high-speed, 67, 131, 136
   iteration, 19, 27
   liquid, 66
     two-dimensional, 73
   multidimensional, 74
   multiplex, 10, 13, 190, 191
   on-column reaction, 105
   steam–solid, 192
   stopped-flow, 20, 27
   tandem system, 92
   time-resolved, 60
   two-dimensional, 21, 59, 74
   vacancy, 19
chromato-mass spectrometer, 153
circulant matrix, 38–40, 49
coal pyrolysis, 82
Coanda effect, 129
cold-trapping, 126
column, alumina, 77
   analytical, 75

Index

modulator, 119
open-tubular, 65, 168
Partisil 5 and 10, 78
polarity, 90
silica, 77
switching, 73–74
  automatic, 76
  in gas chromatography, 79–84
  in liquid chromatography, 76–79
complex injection, 16
computer, Apple, 94, 137, 171
  demands of multiple-input, 136, 137
  graphics, 94
  interface, 137
  micro, 17
  personal, 17, 72, 137
  pneumatic, 137
  tomography, 101
concentration impulse, 15
continuous analysis, 21
  chromatograph, 21
  flow reactor, 104
contour plot, 96, 97, 177
  BASIC subroutine, 215, 216
convolution integral, 15, 16, 22, 26, 28, 32, 38, 54
correlation chromatography, 28, 56
  frequency-modulated, 187
  gas chromatography, 185
  history, 186–189
  HPLC, 189, 193, 194
  in field measurements, 189, 193
  in gas analysis, 189
  in space studies, 192
  in water analysis, 192–196
  resolution, 55
  signal-to-noise ratio, 42
  software, 209
  theory, 37–57
correlation function, 22
  NMR, 33, 34
  noise, 42, 49, 53, 68–70, 102, 125, 189
    mean, 42
    standard deviation, 44, 51, 69
correlogram, 40
cross correlation, 23, 25, 26, 32
crude oil, 86
cryogenic trap, 81, 126
cubic spline approximation, 61
  interpolation, 62
Curie-point pyrolysis, 105–107, 110, 111, 113, 114
  mass spectrometry, 144
  temperature, 110, 111, 114

data processing, 9
Deans switch, 80, 122, 168
decorrelation, 40, 68
  BASIC subroutine, 214
  by direct formula, 41
  by fast Fourier transform, 40–42
degradation, 108
  of acrylic acid–acrylonitrile copolymer, 110
  of isoprene–styrene copolymer, 110

  of polystyrene, 110
  temperature ranges, 164
delta function, 24, 25
demodulation, 21
depth-resolved spectra, 34
desorption of gases from solids, 184
detector, dynamic range, 137
  flame-ionization, 67
  fluorimetric, 193
  noise, 22, 136
    suppression, 22
  output, 30
  response factor, 17
  semiconductor, 190
  time constant, 67, 136
  volume, 67
  ultraviolet, 193
dibenzo-p-dioxins, chlorinated, 76
diesel exhaust, 79
  fuel, 86
differential input, 27
  scanning calorimetry, 140, 150
  thermal analysis, 140, 150
diffusion cell, 103, 104
digital-to-analogue converter, 137
digitization error, 61
  interval, 25, 51
  time, finite length, 64
p-di-isopropylbenzene, 191
drugs in blood plasma, 79

encoding mask, 35
end cutting, 74
ensemble averaging, 131
  correlation, 70–73
entropy, 87
equi-interval sampling, 19, 20, 27, 64, 69, 163
error integral, 18
ethanol in methanol, 82, 83, 86
ethene–ethyl acrylate copolymer, 183
ethene in nitrogen, 189
ethylene–methyl methacrylate copolymer, 158
evolved gas analysis, 139–141, 154–156, 165, 166, 180
exchange column, 19
experiment control program, 212, 213
  design, 10
exponential dilution flask, 102, 103
extra-column band broadening, 29
  effects, 66

factor analysis, 10
fast Fourier transform, 26, 30, 205
fast Hadamard transform, 26, 33, 41, 202, 206, 207, 212, 213
  BASIC subroutine, 214
fast scan NMR, 33
feedback shift register, 42, 202–208
ferromagnetic alloys, 110, 114
  Curie-point temperature, 110, 111, 114
flame retardants, 178, 181
flammability, 142

flow, analytical, 19
  effect of chemical factors on, 119
  effect of physical factors on, 119
  reference, 19
flow modulation, 101, 119–122, 189
  modulators, catalytic, 122
    electrochemical, 122
    photochemical, 122
    thermal decomposition, 121
    thermal degradation, 123
    thermal desorption, 120, 121
      from gallium oxide, 184
      from indium oxide, 184
      of methanol, 184–186
  switch, Deans type, 122–128
    fluidic, 67, 129–131, 135
    pneumatic, 122, 125, 135
    valveless, 125
fluidic switch, 67, 129–131, 135
fluidics, 129
fluorescence, 34
fluorproquazone in animal feed, 78
foreflushing, 126
formaldehyde resin, 185
Fourier transform, 61
  fast, 30
  infrared, 30, 140, 147–150
  ion cyclotron resonance, 30, 34
  ion mobility, 30
  NMR, 30
  spectroscopy, 10, 29, 31
fragrance mixture, analysis, 86
frequency correlation, 163, 189
frequency-modulated input, 27
front cutting, 74
furnace pyrolyser, 106, 111

gas analysis, 189
  chromatography Fourier transform infrared
    spectrometry, 150
    high-speed, 65, 163, 164
    infrared spectrometry, 59
    mass spectrometry, 59
    pyrolysis, 85
  hold-up time, 93
gasoline, 86
geopolymers, 153
ghost peaks, 32, 33
Gidding's formula for peak capacity, 92
Golay equation, 67
greenhouse effect, 190
group separation, 76

Hadamard imaging spectrometers, 36
  masks, 35
  matrix, 39, 41
  transform spectroscopy, 32, 35
    absorption, 37
    infrared, 36, 37
    mass, 35–37
    NMR, 33, 37
    optical, 35, 37
    particle beam, 35
    photoacoustic, 35, 37
    photothermal deflection, 36
    X-ray, 36, 37
    telescope, 36
halogenated pesticides, 86
head-space sampler, 101
heart cutting, 74, 126
height equivalent to a theoretical plate (HETP), 67, 88
herbicides in cereals, 79
hexanol, 85
hidden line removal, 95
high-frequency noise, 42, 44
high-speed separations, 65–67
hold-up time, 20
Huber formula for peak capacity, 92
hydrocarbons, C_1–C_3, 82
  C_1–C_5, 84
  in gasoline and diesel, 86
hydrogen in helium, 189
hydrolysis, 140
hydroxyproline in meat, 79

impulse function, infinitely narrow, 15
  reactor, 104, 105
  response, 15, 32, 33
  technique, 18
inert gases, 86
information, amount, 10, 84ff
  definition, 87
  in two-dimensional systems, 90–92
  theory, 9, 10, 85
injection, complex, 16
  intervals, 165
  multiple, 16
  time, 29
  volume, 165
input function, 16, 26, 27, 118
  iteration, 27
  matrix, 38
    inversion, 40–42
  non-stationary, 48
  sequence, 29, 37
    linear variations, 52, 68
    periodic, 39
    pseudo random binary (PRBS), 39
    random variations, 49
    systematic variations, 51
  sine wave, 118
  single-impulse, 18, 32
  sinusoidal, 21, 22, 27, 41
  stopped-flow, 27
  systems, 118, 134–136
input-output relationship, 15–17, 23, 37, 38, 40
instrumentation, 100
integral transform, 32
inverse matrix, 30, 32, 38
ion cyclotron resonance spectroscopy, 37
ion mobility spectroscopy, 33
isometric projection, 94, 95
isoprene rubber, 166, 167, 173, 177

Index

isoprene-styrene copolymer, 112
iteration input, 27

Kotelnikov theorem, 60

laboratory robots, 65
large samples, 54
least significant bit, 48
linear algebra, 30, 41
   drift, 44
   equations system, 16, 22, 30
   temperature programming, 117
live switch, 126
lanozolac in human plasma, 79

mass spectrometer, 140
   throughput, 68
mathematical processor, 72
matrix, 198
   addition, 198
   algebra, 13, 198
   circulant, 201, 203
   elements, 198
   equation, 30
   Hadamard, 32, 39, 204, 206
   identity, 199
   input, 203
   inversion, 26, 41, 200
   multiplication, 32, 40, 199
   permutation, 200, 203
   subtraction, 199
   Toeplitz, 50
   transposition, 198
   unity, 199
maximum sampling interval, 63
mechanical valves, 67, 79–81, 131–133, 135, 189
mefloquine in human plasma, 79
methane in ambient air, 122, 190
   in earth's atmosphere, 190
   in helium, 189
methanol, desorption, 185
methylene chloride, PCBs in, 86
metoprolol in human plasma, 79
microprocessor, 81, 134
microreactor types, 105
milk, fat, 77
modulation, selective, 121
modulator column, 119
   electrothermal, 122
   thermal desorption, 120
molecular beam, 32, 34
multichannel photodiode arrays, 95
multi-chromatography, 92
multidimensional chromatography, 10, 13
   data, 10
multiple injection, 16
   input, 19, 32
multiple-input computerized chromatography, 100, 139, 170
multiplex advantage, 28, 36, 42, 44, 53, 56, 71, 189, 191
   instrument, 28, 29

   measurements, 28, 29, 32
   NMR, 33
   spectroscopy, optical, 29
multiplicative noise, 42
multivariate statistics, 10
MUSIC (multiple switching intelligent controller), 82, 96, 126–128, 135

naphtha volatiles, 86
natural gas, 86
neutron scattering spectroscopy, 32
nitrile rubber, 166
nitrosamines, 86
nitropolyaromatic hydrocarbons, 79
NMR spectroscopy, 96
   correlation, 33, 34, 37
   fast scan, 33
   Hadamard transform, 37
   multiplex, 33
   PRBS, 33
   pulse, 33
   stochastic, 33, 37
noise, additive, 23
   generator, 24
   pseudo random binary, 25
   standard deviation, 21, 30
   white, 23–25
non-stationary input, 48
normal distribution, 23
notation, 217
nucleic acid bases, 19
null instrument, 19
number of theoretical plates, 66, 91
Nyquist sampling theorem, 60

octanal, 85
on-column reactions, 72, 73
optical correlation spectroscopy, 37
orange extract, 86
organic trace analysis, 79
organochlorine pesticides, 77
organophosphorus pesticides, 77
output matrix, 68, 69
   sequence, decorrelation, 40
   vector, 38, 69
overpressured injection system, 118
oxidation, 140
oxygen in helium, 189

particle beam Hadamard-transform spectroscopy, 35
partition coefficient, 21
pattern recognition, 10
PCBs in methylene chloride, 86
peak amplitude, 55
   capacity, 87–90, 92
      Gidding's formula, 92
      Huber's formula, 92
   dispersion, 25
   forms, 63
   half-width, 17, 63, 64
   mathematical resolution, 66

standard deviation, 17, 64, 66, 67, 69, 88
variance, 25, 54, 66
2-pentanol, 85
periodic reactor, 104
permeation tube, 104
permutation generation, 209
pesticides in milk, 78
  halogenated, 86
  organophosphorus, 86
phenanthrenes in crude oil, 86
phenols in water, 192, 193
phosphonitrilimide, 181, 184
photoacoustic imaging, 35
  spectroscopy, 34
photochemistry, 34
photodiode array detector, 36
photolysis, 140
photolytic reactions, 34
planetary atmosphere studies, 192
plotting two-dimensional chromatograms, 94–97
pneumatic flow switch, 67, 80, 122, 125, 126, 129, 189
polymer characterization, 142
  degradation, of ethylene–methyl methacrylate copolymer, 158
  of polyacrylonitrile, 157, 159
  of polybutadienes, 159
  of polychlorinated biphenols, 86
  of polyether urethane, 148, 149
  of polyimide, 185–188, 192
  of polymethacrylonitrile, 157
  of poly(methyl methacrylate), 148, 149, 157
  of poly(methylvinylpyridine metaphosphate), 181, 184
  of polynuclear aromatics, 79
  of polystyrene, 86, 144–146, 152
  of poly(vinyl chloride), 171
  of poly(vinyl chloride)–poly(vinyl acetate) copolymer, 153
  of urea melamine formaldehyde, 185
polymers, high-temperature resistant, 185
  synthetic, 153
  thermal stability, effect of catalysts, 185
  thermostable, 70, 108, 185
polynomial interpolation, 62
PRBS, *see* pseudo random binary sequence
precolumn, 75
probability, 85
process monitoring, 19
  time resolution, 60
propene in nitrogen, 189
pseudo random binary sequence (PRBS), 25, 29, 30, 32, 35, 38, 39, 103, 166, 203, 212
  algorithm, 42
  noise, 25, 26, 29, 70
pyrolysis gas chromatography, 85, 107, 140–142, 153, 161
  applications, 155
  drawbacks, 155
  historical remarks, 151–153
  impulse, 117
  methods, 153–156

multiple-input, 157–162
one-step two-shot, 161
reproducibility, 108
single-input, 156, 157
stepwise, 157
time-resolved, 171
pyrolysis infrared spectrometry, 147
pyrolysis mass spectrometry, 140, 142–147, 153
  conditions, 144, 145
  Curie point, 144
pyrolysis of biomaterials, 86, 144
  of coal, 82
pyrolysis product analysis, 86
pyrolysis reactors, 105, 106
  arc, 106
  continuous mode, 105
  Curie point, 105–107, 111, 113–115
  electric discharge, 105, 106
  filament, 105, 110, 111, 113, 114
  furnace, 106, 111
  laser, 105, 106, 144
  pulse mode, 105, 107
  radiation-induced, 105
  sample preparation, 106
  sequential, 159
  temperature control, 107, 118
    distribution, 116
    profile, 107, 112, 114, 161, 162
    rise time, 108–112
pyrolysis time, 110

quantization level, 47
  noise, 47
quasicontinuous analysis, 60, 68
quasistationary regime, 117

radiolysis, 140
random input, 22, 25, 27–29
  sequence, 33, 39
reaction chromatogram, 20
  gas chromatography, 20
reactor, 72, *see also* pyrolysis reactors
  chemical micro, 104
  continuous flow, 104, 105
  desorption, 105
  impulse, 104, 105
  periodic, 104
  thermal degradation, 105
refinery gas, 86
resolution, 54, 55
response factor, 51
  function, 19
  system impulse, 15
retention time, 66, 88
roast effect, 113
robotic sampling, 133–135
rubber pyrolysis, 160
  butadiene acrylonitrile, 167, 177
  butadiene methylstyrene, 176, 177
  chloroprene, 166, 167, 177
  degradation products, 175
  studies, 173

Index

ethylene–propylene, 174
fluorinated, 173
isoprene, 166, 167, 173, 177
polybutadiene, 159, 176, 177
silicone, 173
styrene–butadiene, 174
rubber working temperature, 173, 174

sample concentration, by adsorption, 55
  drawbacks, 55
  preparation for pyrolysis reactors, 106
  sources, 100, 101
    diffusion cell, 103, 104
    exponential dilution flask, 102, 103
    permeation tube, 104
  trapping, 78
  variations, 64
sampling, equi-interval, 64
  head-space, 101
  interval, 54, 60, 63, 65, 68, 166, 167
  number of samples, 63
  pseudo random, 69
  stroboscopic, 64, 72, 73
  theorem, 25, 60–64, 69, 75, 157
  time, 67, 135
selectivity tuning, 92–94
senna glucosite, 79
separation quality function, 94
Shannon formula, 87
SiChromat 2 chromatograph, 81, 82, 86, 126, 135
signal-to-noise ratio, 32, 34, 42, 44, 52, 69, 103
simplex optimization, 94
sine wave input, 118
single-impulse injection, 16
  input, 18, 32
sinusoidal input, 21, 22, 27, 41
spatial multiplexing, 34
spectra, depth-resolved, 34
spectrometric imager, 36
Spherosil, 192
spikes, 44, 46
stack plots, 94, 95, 177
step input, 18, 27
  technique, 18
stochastic excitation NMR, 33
stopped-flow chromatography, 20
  input, 27
stroboscopic sampling, 64, 72, 73
structural properties, 34
styrene impurities, 86
superposition principle, 16
switching valves, 25
system impulse response function, 15

tandem system, 92, 93
Taylor series, 52
temperature control, 107, 118
  distribution, 116
  profile, 109
  rise time, 108, 109
tetrabutaline in human plasma, 79
theoretical plate,

thermal analysis, 70, 140, 150
  conductivity detector, 19
  decomposition, 121
  degradation, 140
    characterization, 140
  modulator, 123
  products, 69, 70
  properties, 34
  desorption modulator, 120
  wave imaging, 34, 37
thermochromatography (ThGC), 65, 84, 162–166
  apparatus, 169
  applications, 171
  computer software, 171
thermogravimetry, 65, 140, 141, 150, 151, 163
thermomechanical analysis, 140, 150
thermostability, 142, 150
thermostable polymers, 70, 108
time of flight spectroscopy, 32, 35, 37
time-resolved pyrolysis, 145, 171
  selectivity, 171, 172
time-varying flow, 60
  processes, 68–70
tobacco smoke, 86
Toeplitz matrix, 50
transform techniques, 32
truncation error, 64
two-dimensional chromatograms, measurements, 59, 60
  presentation, 94–96
    contour plots, 96, 97
    slices, 95–97
    stack plots (isometric projection), 94, 95, 97
two-dimensional NMR, 59, 60
two-dimensional systems, comparison, 82

uncertainty and information, 84–87
UNIVAP, 82, 86
urine, 78

vacancy chromatography, 19
valveless flow switches, 125
valves, input, 27
  mechanical, 67, 79, 80, 131–133, 135
  solenoid, 80, 132, 133
variance, 29, 30
vinyl polymers, 152
vitamin B in food, 79
Volterra series, 17

water analysis, 192
waveform of input system, 118
weighing design, 29, 35
white noise, 23–25
Wiener-Lee relationship, 23
wine volatiles, 86, 91

Zymate robot, 133